Thomas Schulenberg (Ed.)

Annual Report 2016 of the Institute for Nuclear and Energy Technologies

Karlsruhe Institute of Technology

KIT SCIENTIFIC REPORTS 7742

Annual Report 2016 of the Institute for Nuclear and Energy Technologies

Edited by
Thomas Schulenberg

KIT Scientific Publishing

Report-Nr. KIT-SR 7742

Impressum

KIT Scientific Publishing

Karlsruher Institut für Technologie (KIT)
KIT Scientific Publishing
Straße am Forum 2
D-76131 Karlsruhe

KIT Scientific Publishing is a registered trademark
of Karlsruhe Institute of Technology.
Reprint using the book cover is not allowed.

www.ksp.kit.edu

Print on Demand 2017 – Gedruckt auf FSC-zertifiziertem Papier

ISSN 1869-9669
ISBN 978-3-7315-0706-2
DOI 10.5445/KSP/1000072764

Content

Institute for Nuclear and Energy Technologies

Structure and Activities of the Institute for Nuclear and Energy Technologies

Thomas Schulenberg

Mission

The Institute for Nuclear and Energy Technologies (Institut für Kern- und Energietechnik, IKET) is situated with its offices and research laboratories on the North Campus of KIT. It is focused on nuclear, fusion and renewable energy technologies for electric power production and on hydrogen technologies for energy storage in chemical form. Its research topics include analyses and tests of thermal-hydraulic phenomena, combustion phenomena and neutron physics which are typical for normal operation or for accidental conditions in nuclear power plants, for future nuclear fusions reactors, for geothermal or solar power plants, but also for mobile systems. Most subjects are application oriented, supported by some basic research projects, if needed.

Organizational structure

IKET is structured into nine working groups as indicated in Fig. 1. Working groups on Accident Analyses, on Accidents Management Systems and on Transmutation as well as the AREVA Nuclear Professional School have been concentrating in 2016 primarily on nuclear applications, whereas the Karlsruhe Liquid Metal Laboratory (KALLA) and the Hydrogen group were addressing nuclear as well as renewable energy technologies. The working group on Magneto-Hydrodynamics is primarily working on nuclear fusion applications, whereas the working group on Energy and Process Engineering is rather concentrating on geothermal energies. Thus, the institute covers a wide field of different energy technologies, and the share

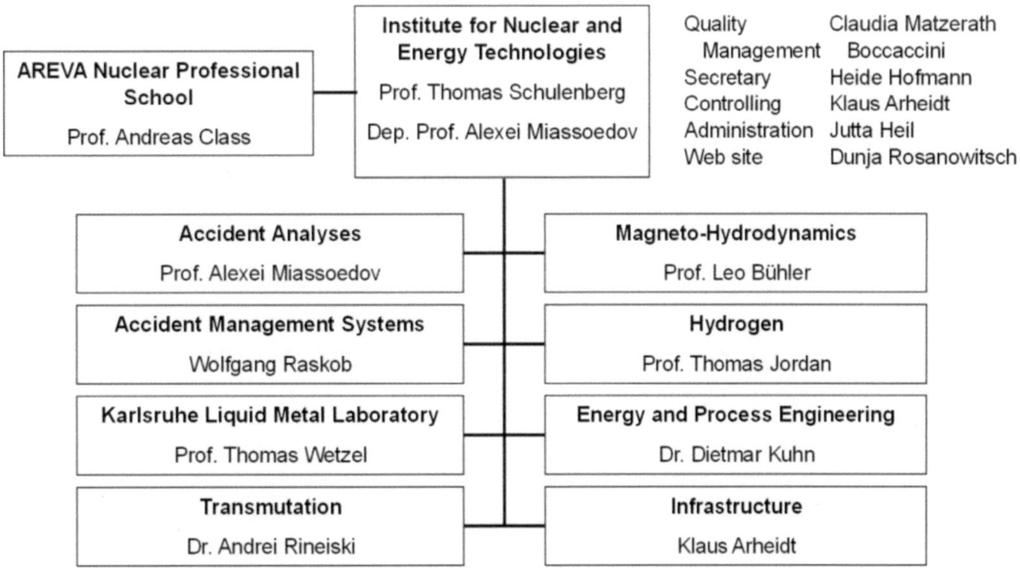

Fig. 1: Organizational structure of the Institute for Nuclear and Energy Technologies (IKET)

of its personnel resources on the different research topics is determined each year by the worldwide market request for energy research.

By the end of 2016, IKET had employed more than 110 (full time equivalent) scientists, engineers, technicians and other personnel. Around half of the employees were funded in 2016 by the Helmholtz Association (HGF), the others by third party funds of the European Commission, by industry, by German ministries or by other research funds. Doctoral students as well as students of the Baden-Wuerttemberg Cooperative State University (DHBW) were filling around 20% of these positions at IKET. In addition, students perform their bachelor or master theses or spend an internship in the research laboratories of IKET.

IKET contributed in 2016 to the HGF programs “Nuclear Waste Management and Safety (NUSAFE)”, “Renewable Energies (RE)”, “Storage Systems and Cross-Linked Infrastructures (SCI)”, as well as “Nuclear Fusion (FUSION)”. More than 2/3 of all IKET personnel was needed to solve issues on nuclear safety, on severe accident research or on alternatives for nuclear waste management.

International collaboration

Most studies performed at IKET are embedded in international research projects. In nuclear safety research, A. Miassoedov is coordinating the European project SAFEST on severe accident research, addressing European safety targets, as well as the European project ALISA, providing transnational access to large infrastructures in Europe and China for severe accident research. Moreover, his research group at IKET is contributing to the European project IVMR on in-vessel melt retention for severe accident mitigation, to the European project FASTNET, predicting the potential radioactive source-term during nuclear accidents, as well as to a collaboration with French research organizations on ex-vessel core melt mitigation systems, which may be retrofitted in existing pressurized water reactors.

Regarding the rapid increase of nuclear power in China, where currently 6 to 7 new nuclear reactors are commissioned each year, the transfer of safety technology from Europe to China is highly appreciated on both sides. This is expressed also by the recent appointment of A. Miassoedov as an adjunct professor by Xi’an Jiaotong University in China, building-up a long-term partnership between both research organizations.

Research on emergency management for nuclear accidents is primarily continued in Europe. W. Raskob is coordinating the European project PREPARE, producing integrated tools and platforms for radiological emergency preparedness and post-accident response in Europe. He coordinates also the new European project CONFIDENCE to improve the decision support for nuclear accident management under high uncertainty. The RODOS software, developed by this team, is being extended since 2016 to account also for risks of tornados, flooding and bush fire in the restricted area around Chernobyl. Moreover, China decided to customize the real-time online decision support system RODOS to Chinese conditions. Supported by the European Commission, IKET installed the RODOS system in the Chinese emergency center in Beijing.

Though liquid metal cooled nuclear energy systems are currently not considered in Germany, IKET is contributing to the safety assessment of such innovative systems by participation to the European projects SESAME, dedicated to thermal-hydraulics and safety of liquid metal cooled systems, and to the European project ESNII+ improving the safety performance of sodium-cooled fast reactors. Moreover, by participation in OECD and IAEA projects on liquid metal cooled reactors, scientists at IKET are self-committed to keep their expertise in nuclear engineering even outside the frame of German energy policy. Their expertise is acknowledged also by utilities and regulators in France and Japan, confirming a high quality of nuclear research at IKET.

International collaboration in nuclear fusion is organized by ITER, F4E and EUROfusion, to which the Magneto-Hydrodynamics Group of IKET is contributing with experimental and numerical studies on liquid metal flow in blankets for fusion reactors.

Outside nuclear fusion and fission research, O. Jedicke at IKET is coordinating the European projects H2FC on the European infrastructure supporting hydrogen and fuel cell technology, and the European project SUSANA, aiming at verification and validation of CFD software for analyses of hydrogen risks. In 2016, a new European project NET-tools has been granted, with the objective of novel education and training tools, based on digital applications related to hydrogen and fuel cell technologies. This project, again, is coordinated by O. Jedicke, IKET.

be expected on the tube surfaces has been measured in 2016 with a new method, confirming the appropriateness of the design. The liquid metal loop of the solar receiver shall be completed in 2017.

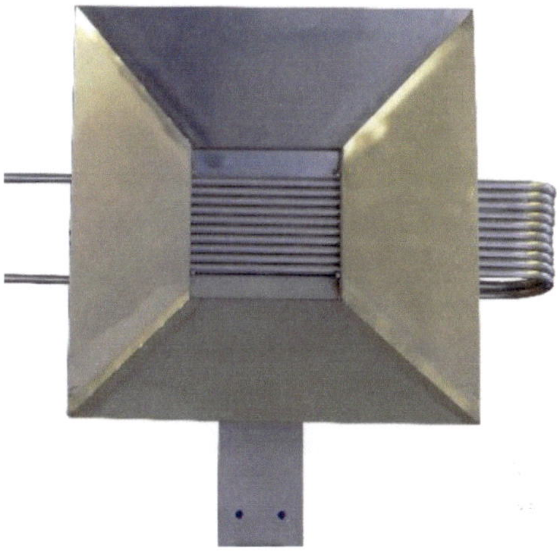

Fig. 2: Liquid metal solar receiver of the concentrated solar power facility SOMMER of the KALLA lab.

Some research highlights of 2016

Out of the large number of research results in 2016, a few examples shall be highlighted here to illustrate the wide scope of research fields at IKET.

Liquid metals are offering an excellent potential as heat carrier in concentrated solar power plants: With their high boiling point, the solar receiver can be operated at high temperature, enabling a better thermal efficiency of the power plant. The good thermal conductivity, on the other hand, can homogenize hot spots of the solar receiver, which might arise due to imperfect heliostat mirrors. In its SOMMER test facility, the Karlsruhe Liquid Metal Laboratory KALLA has completed the installation of a planar heliostat mirror, which follows the sun, shining on a parabolic mirror, which concentrates the solar power onto the receiver. The receiver, Fig. 2, is made of parallel stainless steel tubes, which are cooled inside with liquid lead-bismuth. The solar power distribution to

An organic Rankine cycle with supercritical propane is being installed by the Energy and Process Engineering Group of IKET. Compared with conventional, subcritical Rankine cycles, this innovative approach shall enable a heat-up of the working medium with minimum exergy losses, and thus a maximum thermal power from low temperature energy sources. The new test facility MONIKA comprises a heating unit, a geothermal energy source for test, the Rankine cycle including pump, heat exchanger, turbine and condenser, as well as the propane supply unit with gas purification for a thermal cycle power of about 1 MW. In 2016, all major components of MONIKA have been tested individually before delivery. Finally, they were shipped to KIT and installed on the test site. Fig. 3 shows the pump and the hot side heat exchanger of MONIKA. Commissioning of the test facility will follow in 2017.

Fig. 3: Pump and heat exchanger of the organic Rankine cycle MONIKA at IKET.

Already since the 1990s, but in particular since the catastrophic failure of the boiling water reactors in Fukushima, severe accident mitigation for existing nuclear power plants became a top priority research issue of the Accident Analysis Group of IKET. In case of a failure of the reactor pressure vessel, the interaction of the core melt with concrete in the reactor cavity can be mitigated with a cooling structure, comprising a porous concrete layer, which is fed by cooling water from a pool inside the containment, and a dense, sacrificial concrete layer on top. In 2016, such core catcher design has been tested at IKET with more than 1t of thermite and zirconium melt poured onto it to simulate the core melt. The residual heat was simulated by addition of Zr and thermite, which was supplied to the melt during the test. A core melt layer of up to 60 cm height at a temperature of around 1900°C could be cooled successfully with this concept. Fig. 4 shows a cut through the porous concrete, with the cooled oxide and metal layers on top, inside the test crucible.

Fig. 4: Result of a successful core catcher test with 1150 kg of metal and oxide at 1900 °C from a thermite reaction

Severe accident mitigation will be even more challenging for liquid metal cooled reactors. While the sodium cooled reactor ASTRID is being developed in France, aiming at a sustainable nuclear energy supply, the Transmutation Group at IKET is focusing on analyses of potential severe accidents of such reactors. The multi-physics code SIMMER, to which IKET is contributing, simulates a severe accident progression, e.g. in case of an unprotected loss of flow through the reactor core, including reactivity response, thermal and mechanical consequences, and even degradation of the core. In 2016, code applications have been extended to three dimensions, which can simulate flow conditions inside a pool reactor more realistically. SIMMER analyses of a new ASTRID design, Fig. 5, which have been performed in 2016, could confirm that an unprotected loss of flow accident would result in less damage. Long term consequences of such accidents shall be studied in 2017.

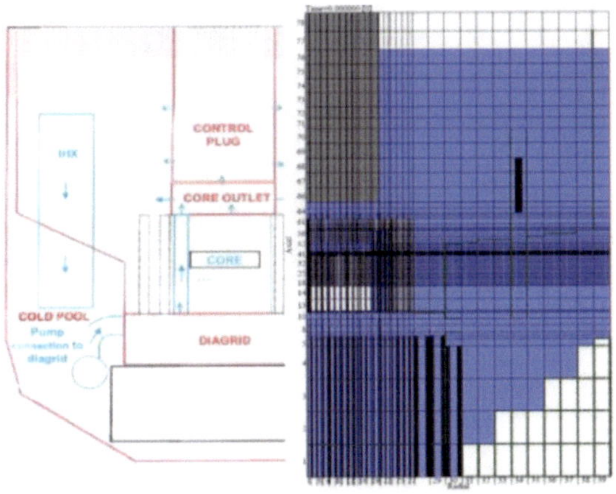

Fig. 5: SIMMER model of the ASTRID primary system.

Besides research performed in the area of nuclear emergency management, the Accident Management Systems Group is active in the area of protection of critical infrastructures (CI). In cooperation with institutes of KIT Campus South, an agent-based simulation system was developed to describe the dependencies of CIs on electricity and water supply for a future smart city. So far, IKET is focusing on health care, and comprehensive models were developed covering hospitals, pharmacies, dialysis centers and retirement homes. The objective of this work is to investigate sensible management options, in particular focusing on optimization processes to keep the performance of the CI on a level that allows the secure supply of the population with all their needed products.

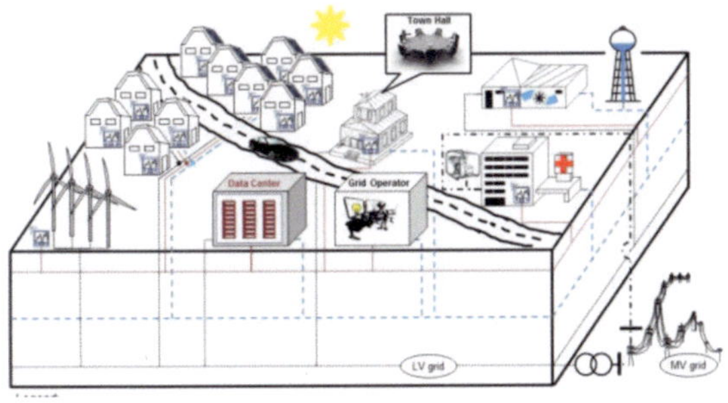

Fig. 6: Key components of critical infrastructures considered in the agent-based simulations.

The AREVA Nuclear Professional School at IKET is committed to education and training of post-graduate students and young professionals in nuclear engineering. Up to 20 courses on science and technology of nuclear reactors have been offered in 2016, including i.a. neutron physics, thermal-hydraulics, stress analyses, accident management and a new course on human-machine-interaction. Five doctorates on nuclear engineering have been completed at IKET in 2016, which were covering numerical predictions of two-phase flow in fuel assemblies and in safety systems, numerical simulations of core melt retention, reactor physics, and magneto-hydrodynamic effects of liquid metal flow in blankets of nuclear fusion reactors. As an example of advanced modelling of two-phase flow, the dissertation of Matthias Benz is using a statistical approach to describe the wavy interface of a stratified flow of water and steam, whereas the dissertation of Matthias Viellieber introduces coarse grid computational fluid dynamics for industrial applications.

The Magneto-Hydrodynamics Group at IKET is focusing on blanket development and test for nuclear fusion reactors. Experiments for measuring the magneto-hydrodynamic pressure drop in a scaled mock-up of the helium cooled lead lithium (HCLL) test blanket module (TBM) for ITER have been performed in the MEKKA facility at IKET. The mock-up consists of 8 breeder units (BUs) connected two by two at the first wall through small openings. Breeder units are fed and drained by a system of manifolds. A photograph of the test section in front of the magnet is shown in Fig. 7. The pressure drop is measured conventionally with pressure taps, whereas velocities can be concluded from electric potential measurements outside the test module.

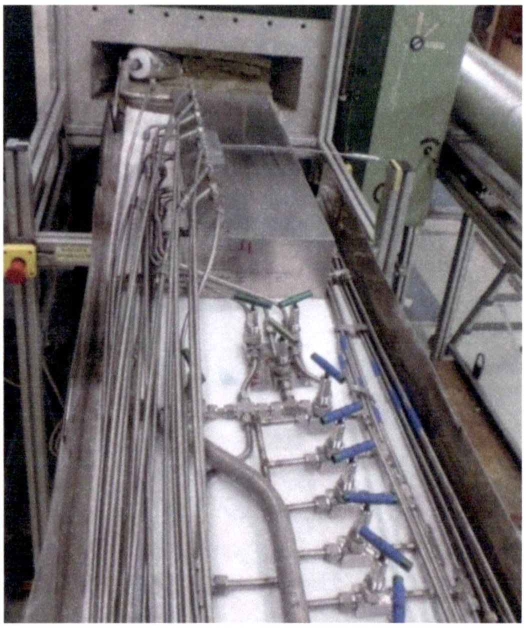

Fig. 7: Test section of a breeder unit being installed in the MEKKA facility.

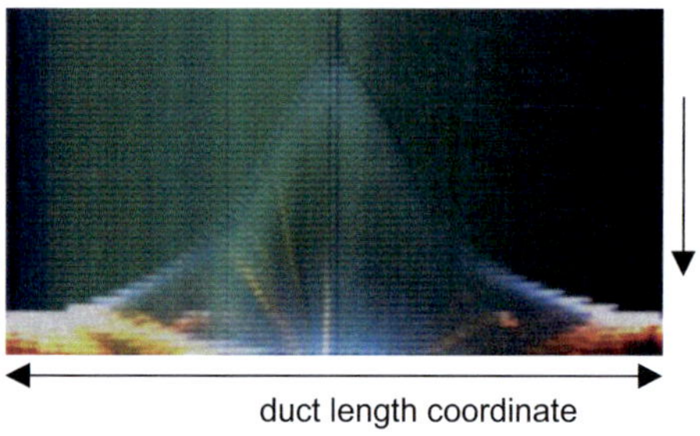

Fig. 8: Symmetric flame front progression in a transparent duct

Besides model and numerical code development the Hydrogen Group of IKET was continuing its experimental work on fundamental dynamic combustion behavior of premixed systems. Experiments for flame acceleration and deflagration-to-detonation-transition in micro channels have been inspired by industry partners, who integrate fuel cells and electrolysers in their systems. They had observed surprisingly fast combustion regimes in these relatively small structures. Complementing the related flat layer experiments executed in 2015, experiments have been conducted in a small but long channel, with transparent walls, open ends and variation of the cross-sectional dimensions.

Fig. 8 shows a sequence of frames taken from high speed videography of this 1.8 meter long and 2 cm high (and deep) transparent duct. Each frame is arranged below the other. Thus, the vertical coordinate coincides with the time coordinate. The illuminated zone indicates the flame front spreading symmetrically to the left and right from the central point of ignition. The angle enclosed by this flame front line and the central vertical symmetry line indicates the flame speed. Short after ignition (top line), we see a stable deflagration with a pale flame, which accelerates only after a certain time to turn finally into a detonation. This deflagration-to-detonation-transition is visible by the strong flashing light emitted in the lower left and right corner of Fig. 8. In general, the stability of flame, effective flame speed and associated pressure developing in such narrow channels were found to be strongly dependent on the cross-sectional area, the congestion and mixing properties. The data generated with these experiments are used for further model and code validation.

Group: Magnetohydrodynamics

Magnetohydrodynamics for Liquid-Metal Blankets

Leo Bühler, Thomas Arlt, Hans-Jörg Brinkmann, Victor Chowdhury, Christina Köhly, Chiara Mistrangelo

Overview of MHD phenomena related to electromagnetic flow coupling

For the development of liquid metal blankets for fusion reactors, the analysis of magnetohydrodynamic (MHD) effects, caused by the interaction of the flowing electrically conducting medium with a strong imposed magnetic field, is of crucial importance. In complex geometrical configurations, electric currents flow from one fluid domain into the adjacent ones passing through common electrically conducting walls. The resulting electromagnetic coupling between parallel ducts modifies velocity and pressure distributions compared to the ones in single channels. In order to complement earlier work [1] [2] [3], we investigate numerically MHD flows of incompressible, electrically conducting, viscous fluids in channels that are coupled via conducting walls. A uniform transverse magnetic field is applied, which is inclined at an arbitrary angle α to the z-axis, as shown in Fig. 1. The flow in the ducts is driven by given pressure gradients that can vary from one channel to another.

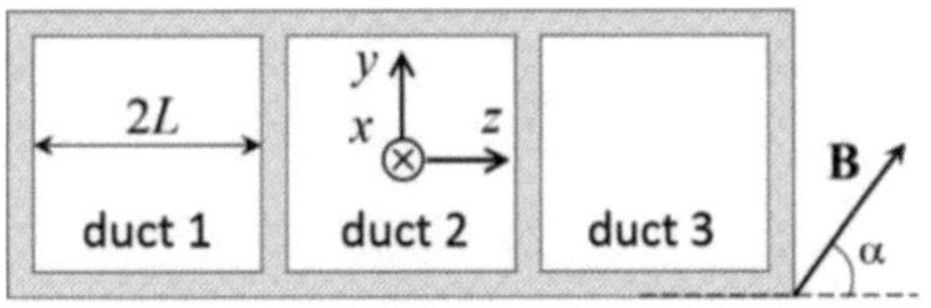

Fig. 1: Electrically coupled channels exposed to an arbitrary oriented magnetic field.

The steady state flow is governed by the dimensionless equations for conservation of momentum, mass, charge, and by Ohm's law:

$$\nabla p = -\hat{\mathbf{x}} = \frac{1}{Ha^2}\nabla^2 \mathbf{v} + \mathbf{j} \times \mathbf{B}, \qquad (1)$$

$$\nabla \cdot \mathbf{v} = 0, \qquad \nabla \cdot \mathbf{j} = 0, \qquad (2)$$

$$\mathbf{j} = -\nabla \phi + \mathbf{v} \times \mathbf{B}. \qquad (3)$$

The non-dimensional parameter in (1) is the Hartmann number $Ha = BL(\sigma/\rho\nu)^{1/2}$. Its square describes the ratio of electromagnetic to viscous forces. The variables $\mathbf{v}$, $\mathbf{j}$, $\mathbf{B}$, and ϕ denote velocity, current density, magnetic field, and electric potential, scaled by u_0, $\sigma u_0 B$, B, and $u_0 BL$, respectively. The typical length L of the problem is the half-width of a duct and the characteristic velocity u_0 is obtained from the given pressure gradient as $-\partial_x p/\sigma B^2$. Density ρ, electric conductivity σ, and kinematic viscosity ν, are constant. Since the walls of the ducts are electrically conducting, equations for electrical variables are solved also in the solid domains.

Results discussed in this section refer to a constant Hartmann number $Ha = 250$. We begin with the description of cases in which the flow in the 3 ducts is driven by the same pressure gradient $\nabla p_{1,2,3} = \nabla p$. We consider first two limiting cases, i.e., when the magnetic field is applied in y-direction ($\alpha = 90°$), and when $\mathbf{B}$ is parallel to z ($\alpha = 0°$) to provide a complete picture of the influence of the inclination of the magnetic field on electromagnetic coupling in ducts.

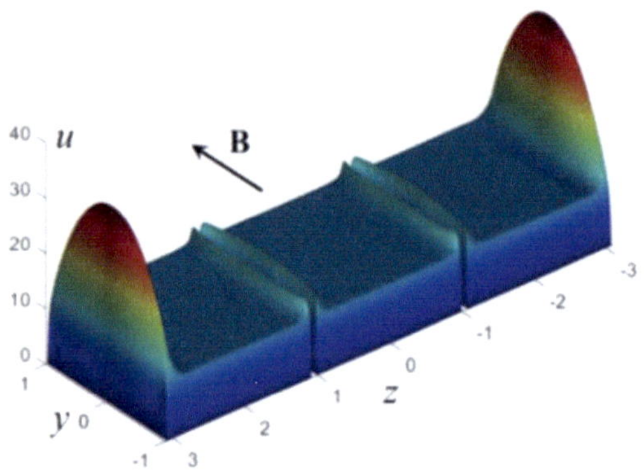

Fig. 2: Velocity distribution for the flow with (α= 90°).

Fig. 2 shows the velocity distribution in three channels, when **B** is parallel to y. Currents induced in the fluid cross the internal separating walls, which behave as if they were almost perfectly conducting. Therefore, in the side layers along dividing walls only a slight increase of velocity compared to the core value is observed. At external sidewalls, current flows preferentially in magnetic field direction and induced Lorentz forces are much smaller than the ones in boundary layers at internal walls. As a result high velocity jets occur along the external walls as expected for MHD flows in electrically conducting ducts. Consequently, the lateral channels carry a larger flow rate than the central one.

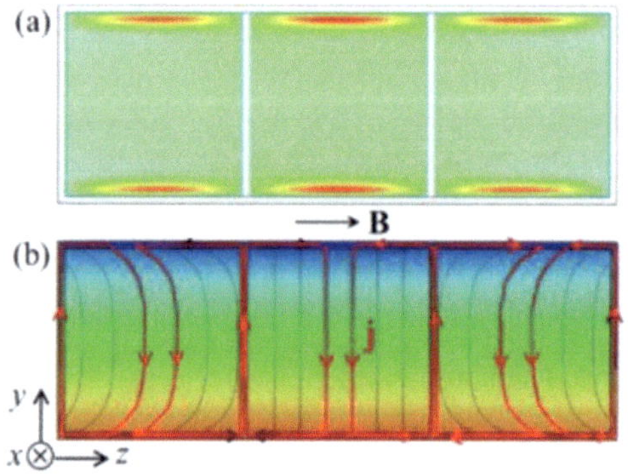

Fig. 3: (a) Velocity and (b) electric potential contours and current streamlines for (α= 0°).

Velocity and electric potential distributions are depicted in Fig. 3 for the case in which the magnetic field is aligned with the z-direction. In layers parallel to **B**, jets with high velocity are present and the core flow in the central duct is uniform. Current streamlines in duct 2 are symmetric with respect to the center of the duct, while in channel 1 and 3 they are asymmetric. In this case, with α = 0° and equal driving pressure gradients, there is no exchange of current through common Hartmann walls, perpendicular to **B**, and the influence of electromagnetic coupling appears weak.

Let us consider now the case, when the magnetic field is inclined at an angle α = 15° to the z-direction. Fig. 4 displays typical electric current paths together with contours of electric potential (a) and axial velocity (b). Potential contours show that, as a result of the electromagnetic coupling of neighboring flow domains, cores exist that extend through the three ducts across the dividing walls. They are separated by internal layers that spread along magnetic field lines starting from duct corners. Along magnetic field direction, current density is constant, electric potential varies linearly, and according to Eq. (3) also the axial velocity has a linear variation.

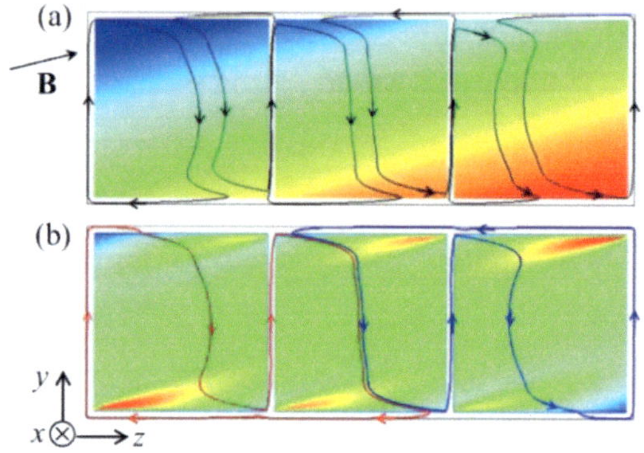

Fig. 4: Typical current loops for α= 15°. Contours of (a) electric potential, (b) axial velocity.

In the internal layers the velocity increases when moving towards the channel corners, as visible in Fig. 4(b) and in Fig. 5.

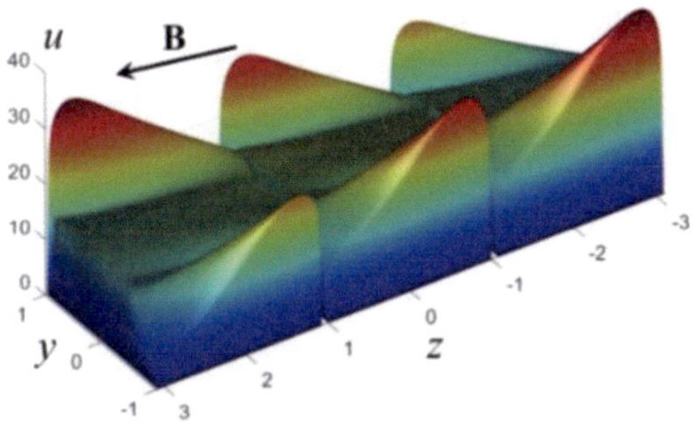

Fig. 5: Velocity distribution for the case with α= 15°.

Two types of characteristic current paths can be identified for this flow (see Fig. 4): those that close in a single duct (a) and the ones that couple two channels (b). The second kind of loops that connects two ducts is such that leakage currents, before turning in the adjacent channel, flow either along the separating walls or along top and bottom walls.

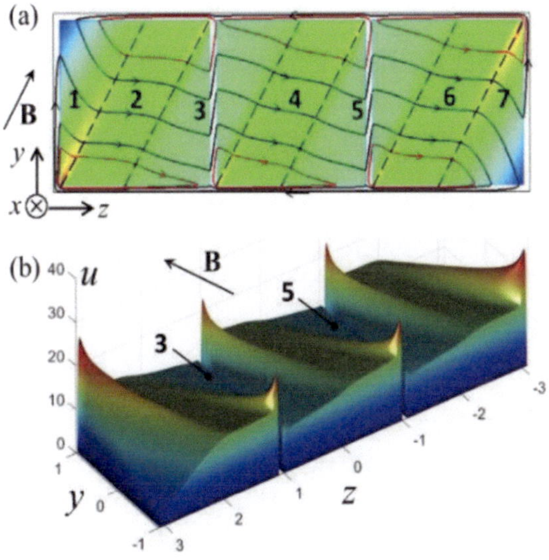

Fig. 6: (a) Typical current loops and contours of axial velocity, (b) velocity distribution for the flow with α= 60°. Cores are marked by numbers.

For larger α, the electromagnetic coupling becomes stronger. An example is plotted in Fig. 6 for α = 60°. Current paths marked in black show how the induced current flows from one duct to the next one forming large loops that couple all the three channels. Regions near the corners still exist, from which the internal layers originate, where current closes just in a single duct. In Fig. 6, dashed lines trace the internal layers that split the cross-section in 7 cores. Core 3 as well as core 5 are hydraulically split in two parts by the dividing walls, but they are strongly electrically coupled.

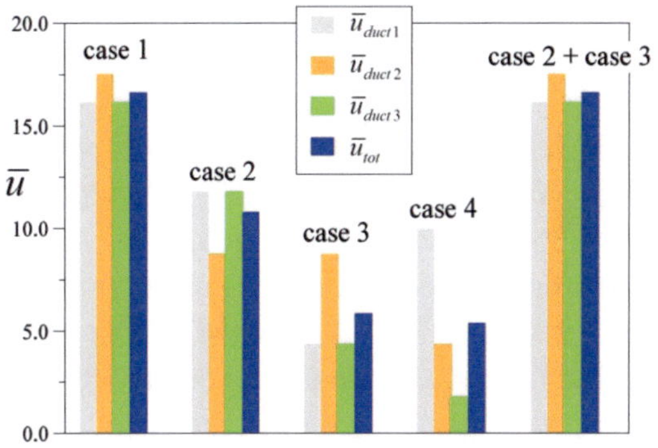

Fig. 7: Average velocity in the ducts and total mean value for α= 15° and 4 cases: 1) $\Delta p_{1,2,3}=\Delta p$, 2) $\Delta p_{1,3}=\Delta p$, $\Delta p_2=0$, 3) $\Delta p_{1,3}=0$, $\Delta p_2=\Delta p$, 4) $\Delta p_1=\Delta p$, $\Delta p_{2,3}=0$.

Effects of electromagnetic coupling on velocity in parallel ducts can be seen when the driving pressure gradient in one or more channels is zero. In Fig. 7 average velocity in individual ducts and in the total cross-section are compared for different combinations of the applied driving pressure gradients. All cases can be derived by combining cases 3 and 4. In case 1, while for the flow with α = 90° the flow rate was larger in lateral ducts, for α = 15° the opposite situation can be found due to the inclination of **B**.

Fig. 8 and Fig. 9 display 3D views of velocity for flows driven only in lateral channels for two orientations of the magnetic field, α = 90° and α = 15°.

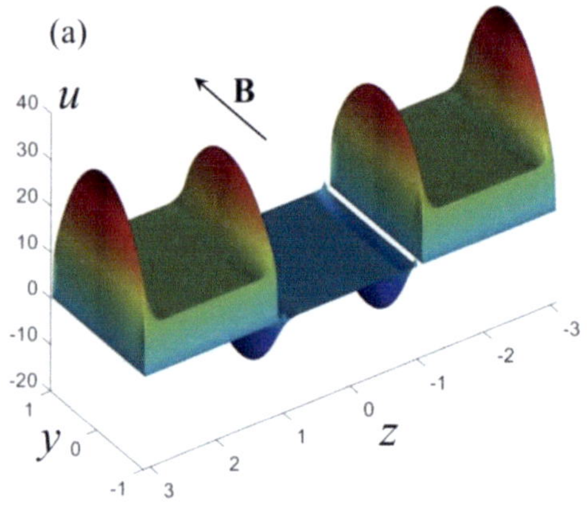

Fig. 8: Velocity distribution in a cross-section for flows driven only in lateral ducts and α= 90°.

In the first case, when ducts are coupled at walls parallel to **B**, core velocity in duct 2 remains zero. Reversed flow occurs in parallel layers. In the second case for α = 15°, ducts 1 and 3 drive the flow in the entire middle channel. Further results and details on electromagnetic flow coupling have been published in [4], [5], [6], [7].

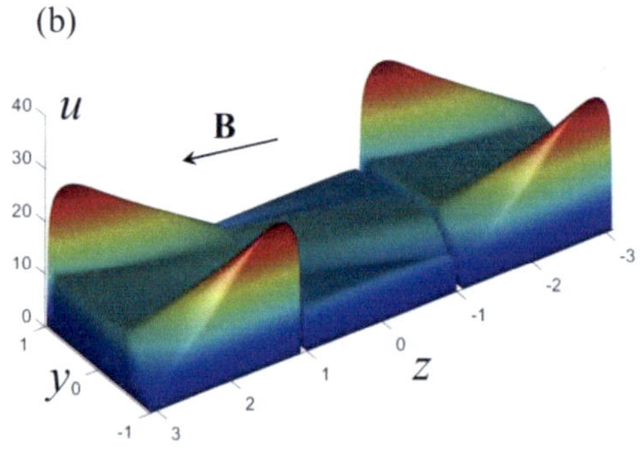

Fig. 9: Velocity distribution in a cross-section for flows driven only in lateral ducts and α= 15°.

MHD flow and heat transfer in model geometries for WCLL blankets

One of the liquid metal blanket concepts considered as option for a DEMOnstration fusion reactor is the water-cooled lead lithium (WCLL) blanket. It consists of modules attached along the poloidal direction to a back supporting structure. Pressurized water cools the walls and breeding zone. In a design concept developed by CEA a large number of cooling pipes arranged in radial-poloidal planes is immersed in the liquid metal which results in a complex flow path for the PbLi [8] (see Fig. 10). Internal plates that form rectangular ducts in which the liquid metal is confined reinforce the breeder zone.

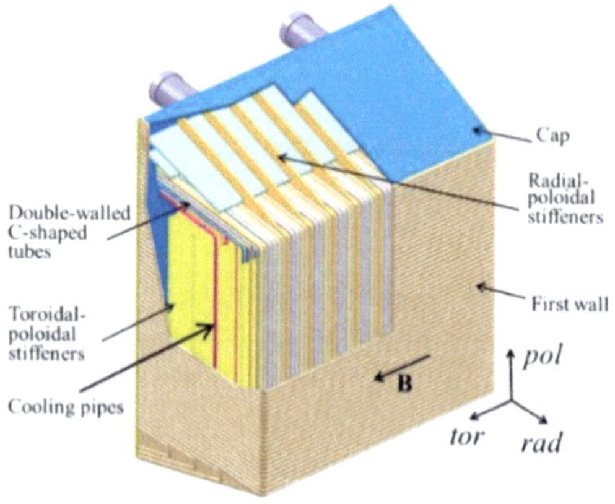

Fig. 10: WCLL blanket according to a design concept developed by CEA.

In order to assess the impact of MHD phenomena on flow distribution and heat transfer in the WCLL blanket concept, numerical investigations are carried out to study MHD flows in generic model geometries. Flows are driven by applied pressure gradients or by buoyancy caused by non-uniform thermal conditions, due to imposed volumetric heat sources and heat removal by cooling pipes.

MHD flows are considered in a WCLL model geometry consisting of an electrically insulating rectangular duct with an internal coaxial cooling pipe (see Fig. 11). The channel is filled with liquid metal heated by a volumetric thermal source. It is assumed that the axial length is long enough that the flow adjusts to fully developed thermal and hydraulic conditions. The model-problem aims at getting insight in details of MHD flows around obstacles such as cooling pipes with heat transfer and buoyancy. A proper understanding of these buoyant-MHD effects is required for interpretation of data of future 3D simulations or upcoming experiments. Numerical simulations have been performed by means of a finite volume code developed based on the software package OpenFOAM [9].

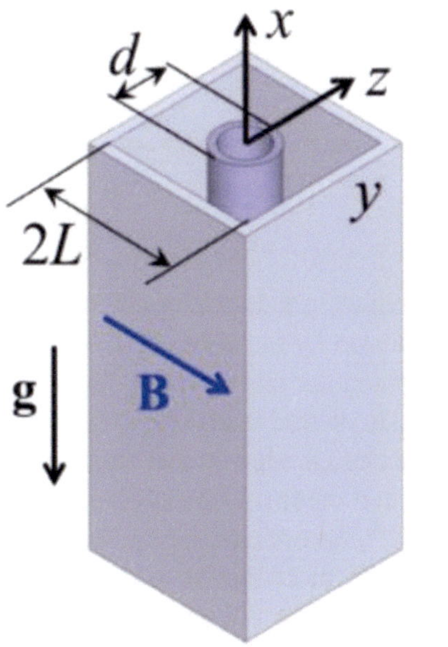

Fig. 11: Sketch of model geometry and coordinates. A circular cooling pipe is inserted into a liquid metal filled rectangular channel.

In a first series of simulations velocity, potential, and current distributions have been determined for pressure driven MHD flows in the model geometry.

Fig. 12 shows contours of potential and streamlines of electric current density for a relatively small Hartmann number $Ha=10$. The topology of current density consists of closed current recirculations near the corners and above and below the pipe. There exists a well-defined region around the pipe from which currents cannot escape.

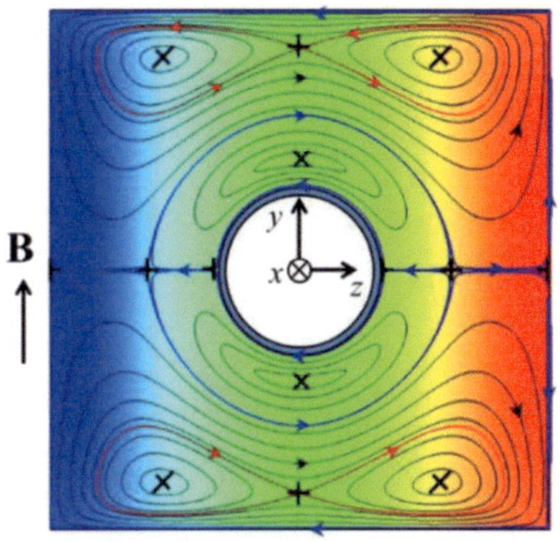

Fig. 12: Contours of potential ϕ and electric current streamlines for pressure driven flow at $\mathrm{Ha}=10$.

A comparison with results for $Ha=1000$ displayed in Fig. 13 shows that the flow develops uniform cores to the right and left (C_{right} and C_{left}) and other cores above and below the pipe (C_{pipe}). The cores are separated from each other by tangent layers of thickness $\delta_t \sim Ha^{-1/2}$ that spread along magnetic field lines. At the right and left duct walls, we find the well-known parallel Shercliff layers of thickness $\delta_s \sim Ha^{-1/2}$. The Hartmann layers with thickness $\delta_H \sim Ha^{-1}$ at Hartmann walls, where the magnetic field has a normal component, are so thin that they are not visible in the figure. Although both figures look quite different, the overall topology remains unchanged. With increasing Ha the critical points move closer to the Hartmann walls and the two saddle points at $y=0$ move into the tangent layer. Closed current recirculations are present above and below the pipe in C_{pipe}, involving roughly half the cores, the tangent layers and the Hartmann layers at the pipe (inside closed blue loop). The four areas enclosed by the red limiting streamlines represent other regions with locally closed current recirculations, mainly located in the outer cores C_{right} and C_{left} with current closure through the Shercliff, Hartmann and tangent layers. Current lines that involve and couple all different cores across the tangent layers are those located between the red and blue loops.

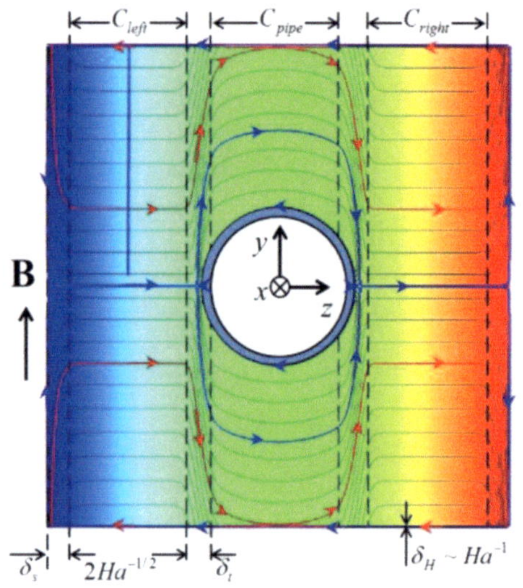

Fig. 13: Contours of potential ϕ and electric current streamlines for pressure driven flow at Ha=1000.

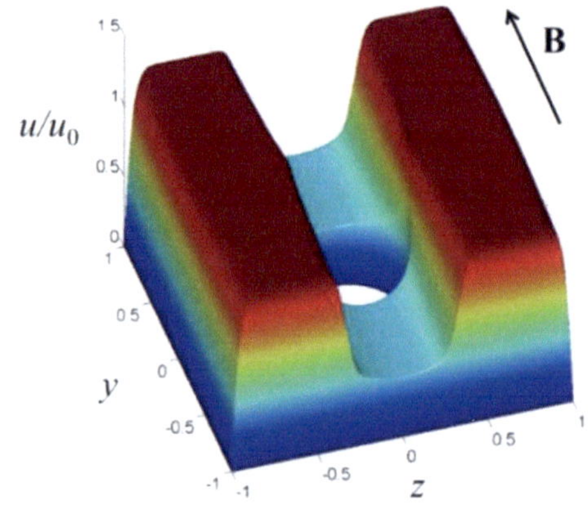

Fig. 15: Axial velocity profile for pressure driven flow at Ha=1000.

Profiles of velocity are shown for Ha=10 and 1000 in Fig. 14 and Fig. 15. For high Hartmann numbers we observe in all cores uniform velocities along magnetic field direction y with thin viscous Hartmann layers at the duct walls and pipe. It is interesting to notice that the velocity profiles in both cases exhibit similar features (apart from the thickness of Hartmann and parallel layers) although the current distributions are significantly different.

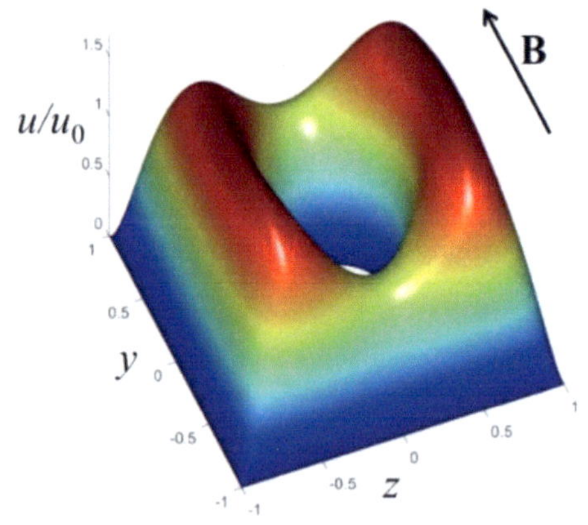

Fig. 14: Axial velocity profile for pressure driven flow at Ha=10.

In the following it is assumed that the flow is purely driven by buoyancy, i.e. we consider a long rectangular closed cavity filled with a volumetrically heated fluid. The latter is cooled by a central circular pipe. In this case, the average velocity in a cross section is zero. At some distance from the ends of the cavity, temperature is independent of the flow. Contours of temperature are shown in the lower half of Fig. 16 for the flow at Ha=10. Around the cooling pipe the isotherms form nearly concentric circles. From a hydrodynamic point of view, it is expected that the cold fluid around the pipe moves downward, while the hot fluid near the outer walls moves upward. This is in fact almost the case for low Hartmann numbers as displayed in Fig. 17.

However, this behavior changes completely for higher magnetic fields. While temperature and resulting buoyancy forces remain unchanged, regions of high downward velocity detach from the pipe and align themselves with magnetic field lines, forming tangent layers (see Fig. 18 for Ha=1000). Even if there exists a radial distribution of temperature, the velocity has a tendency to form uniform cores along y, i.e. the flow exhibits a sort of quasi-2D state. The highest downward velocities are observed in the tangent layers, where the fluid moves down even in regions where buoyancy points upward. Electric current paths become quite complex for higher Ha. We have seen that in pressure driven flows a significant amount of current flows along the tangent layers but for buoyant flows currents just cross these layers. The present model problem might appear as an idealized case due to assumed fully developed conditions for infinite poloidal length. Nevertheless, already for this simple case the flow patterns are quite complex. Further results and details on magneto-convection and flow related to WCLL blankets have been published in [10], [11], [12], [13].

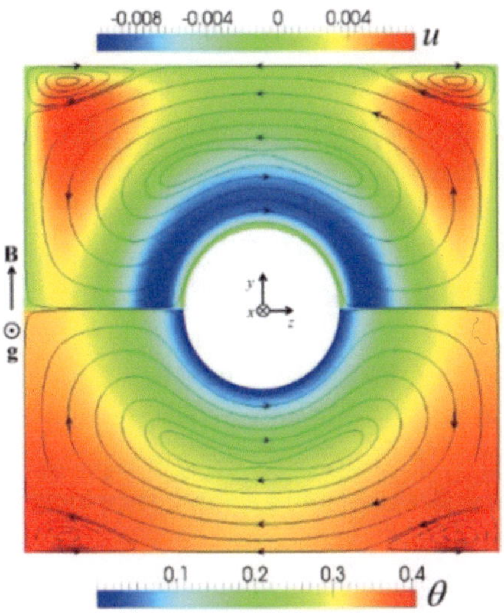

Fig. 16: Electric current streamlines and distribution of temperature (lower half) and of vertical velocity (upper half) for buoyancy driven flow at Ha=10 .

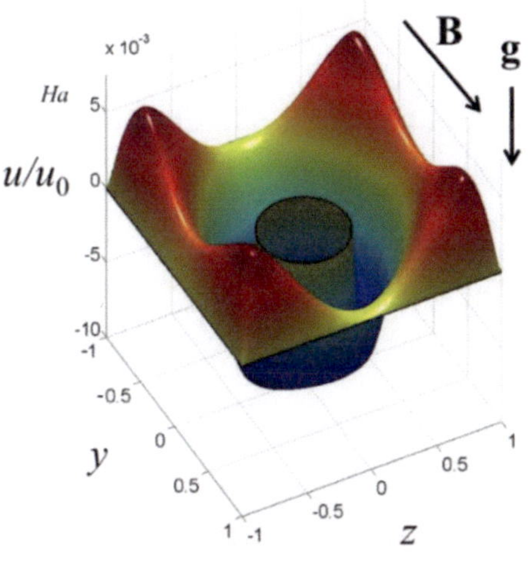

Fig. 17: Axial velocity profile for buoyancy driven flows at Ha=10.

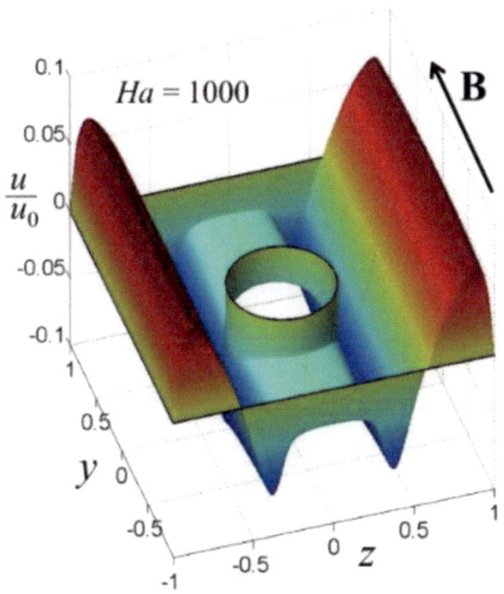

Fig. 18: Axial velocity profile for buoyancy driven flows at Ha=1000.

Pressure distribution in an experimental test-section for a HCLL blanket

In the Helium Cooled Lead Lithium (HCLL) blanket to be tested in ITER the eutectic alloy PbLi is used as breeder material and the generated heat is removed by helium flowing in channels grooved in the walls. In the proposed design concept, a number of rectangular Breeder Units (BU) is arranged in columns to form a blanket module. In the breeder zones a small liquid metal flow rate is required to circulate the PbLi towards external systems for purification and tritium extraction. In the manifolds that feed and drain the BUs PbLi flows with higher velocities due to the smaller cross-sections and because one manifold has to distribute (or collect) the liquid metal into (from) 4 BUs. The interaction of the liquid metal with the magnetic field that confines the fusion plasma, leads to intense MHD effects and higher pressure drop in the manifolds. In order to investigate pressure and velocity distribution in a scaled mock-up, the European HCLL test blanket module for ITER has been fabricated.

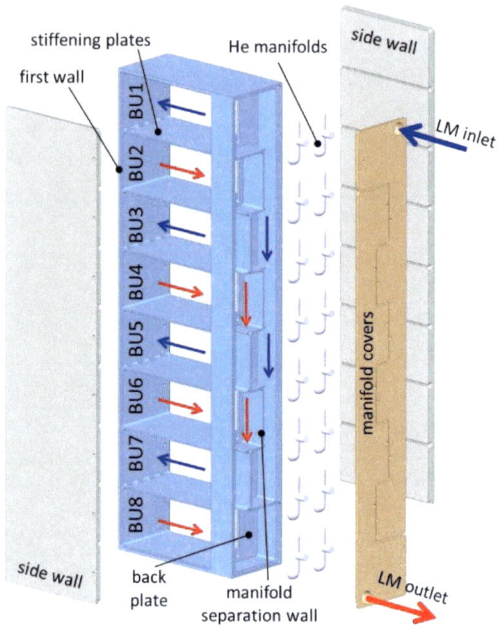

Fig. 19: Exploded view of the MHD mock-up. The liquid metal flow scheme is indicated by arrows. Pairs of BUs are hydraulically connected by means of small opening at the FW.

Experiments have been performed in the MEKKA laboratory of IKET. As working fluid the eutectic alloy NaK, which is liquid at room temperature, is used. Due to its larger electric conductivity and smaller density compared to PbLi used in fusion reactors, it is possible to achieve characteristic flow parameters of the same order as the ones in real applications even if the maximum available magnetic field in laboratory is limited to 2.1 T. An exploded view of the mock-up is displayed in Fig. 19. The main body consists of a single piece, which forms all walls of the BUs, first wall, stiffening plates, back plate and the wall that separates distributing and collecting regions in the manifold. Solid dummy elements simulate partial blockage by helium manifolds. The main dimensions of the scaled mock-up are 798.9×257.5×121.8 mm³. All components are made of austenitic steel that has good compatibility with the working fluid NaK.

Pressure differences between several points on the mock-up have been measured using a piping system that connects the pressure taps to capacitive pressure transducers located out-

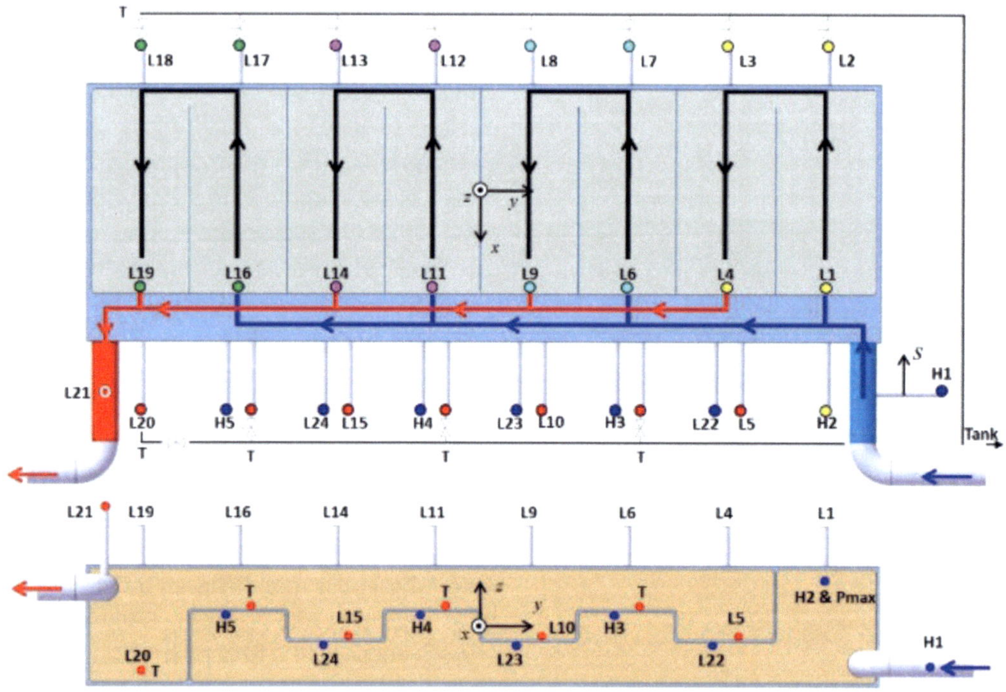

Fig. 20: Sketch showing pressure taps.

side the magnet. The measuring system is designed to measure a pressure difference between lines H_i and lines L_k, both connected to the test section. In the current installation we have i=1,2...5 and k=1,2,....24. The connection scheme is shown in Fig. 20. The pressure distribution is recorded along typical flow paths marked by different colors in Fig. 21, Fig. 22 for various Hartmann numbers Ha and Reynolds numbers Re.

Fig. 21 shows the measured distribution of pressure for Ha=1000 and Re=1014. The dimensional pressure $p*$ is normalized as $p=p*/(\sigma u_0 L B^2)$. Data is plotted along the coordinate s that starts at the pressure tap H1 and ends at L21 at the exit of the draining manifold (cf. Fig. 20). The main contributions to the total pressure drop (H1L21) occur in inlet and outlet circular pipes, (H1H2), (L20L21), in the feeding and draining manifolds (M), (H1H5) and (L5L21), across gaps at the back plate (e.g. H2L1) and at the first wall (e.g. L2L3). At this location the fluid turns in poloidal direction to enter the connected BU by passing across a number of small openings. As a result the velocity increases locally due to the reduction of the cross-section along the liquid metal flow path and inertia forces become stronger. This yields the additional pressure loss. The pressure drop inside breeder units is very small compared to contributions mentioned above, since the liquid metal flows slowly in the large rectangular boxes. It is interesting to observe that the pressure difference between the entrance and the exit of a pair of hydraulically connected BUs is not the same for all couples of BUs. The pressure drop in BU1-BU2 and BU7-BU8 is much larger compared to the one in BU3-BU4 and BU5-BU6. This indicates that there is a higher flow rate in the external pairs of BUs. This is related to the geometry of the manifolds whose cross-section remains constant along the entire path of the liquid metal, while the velocity changes. In the feeding manifold the flow rate reduces progressively and therefore also the pressure drop. In the draining manifold the flow rate increases as well as the pressure loss. A more uniform flow partitioning among breeder units could be only achieved by modifying the design of the manifold so that the average velocity in the manifold keeps a constant value along the liquid metal path.

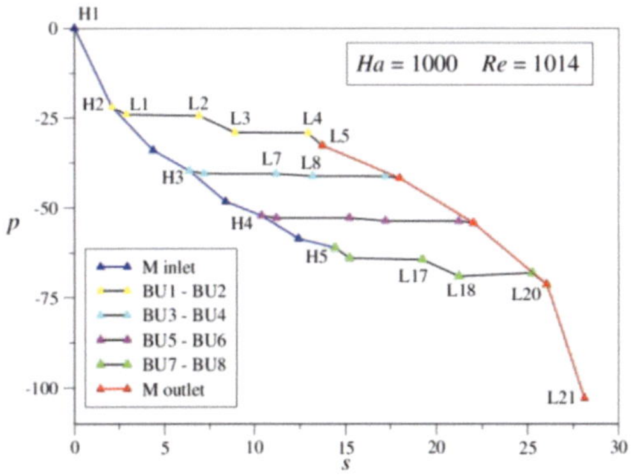

Fig. 21: Pressure distribution along typical flow paths in the blanket module for Ha=1000 and Re=1014. The main contributions to the total pressure drop occur in the manifolds (H1H5, L5L21) and at the first wall in BU12 and BU78.

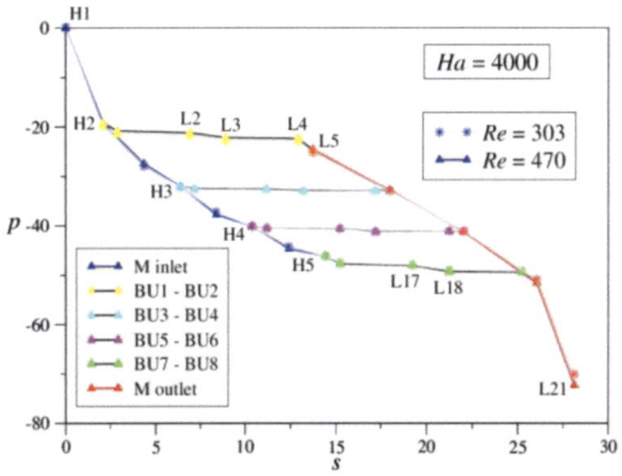

Fig. 22: Pressure distribution along typical flow paths in the blanket module for Ha=4000 and two Reynolds numbers, Re=303, 470.

By increasing the magnetic field strength the influence of inertia on pressure distribution becomes progressively weaker and for $Ha \geq 3000$ nondimensional pressure losses in the mock-up are practically unaffected by the flow rate, even for the highest Reynolds numbers that could be reached in the experiments. An example of pressure distribution at larger Ha is plotted in Fig. 22. Further work in support of HCLL blanket design has been published in [14] [15].

Instabilities in magnetohydrodynamic boundary layers

Magnetically induced instabilities and first occurrence of time-dependent motion has been investigated as contribution to the Helmholtz Alliance LIMTECH. As an example, the stability of Hunt flow [16] is considered since the magnetically induced jets, which develop in duct flows with insulating sidewalls and conducting Hartmann walls, represent a well-defined prototype for potentially unstable MHD flows.

Instability of Hunt-type flow has been investigated experimentally in 2016 for ducts with aspect ratios different from one (previous work). The aspect ratio of the duct was $A = 1/3$ and measurements were performed in a wide range of Hartmann numbers $500 \leq \mathrm{Ha} \leq 2000$ and Reynolds numbers $500 \lesssim \mathrm{Re} \lesssim 37000$. Results for the time-averaged velocity distribution in transverse direction are presented in Fig. 23 for Ha=2000 and various Re. For the smallest Re good agreement between measurements and laminar theory can be observed. The velocity profile becomes non-symmetric for $\mathrm{Re} \geq 5054$. As a result the jet at $z = 0.3$ becomes unstable much earlier, while the other one at $z = -0.3$ does not show any thickening until $\mathrm{Re} = 25782$. A map of flow regimes showing transitions between stable laminar flow, unstable flow with very small fluctuations (unstable 1), and unstable flow with large amplitude (unstable 2) is displayed in Fig. 24. Further results and more details can be found in [17], [18].

First numerical simulations performed at KIT using OpenFOAM for parameters as used in the experiments confirm the onset of instabilities in agreement with predictions of a linear stability analysis [19], [20], [21]. The magnitude of observed perturbations is initially extremely small but grows to significant magnitude for higher Reynolds numbers. Results for other aspect ratios and wall conductivities have been published in [22], [23], [24].

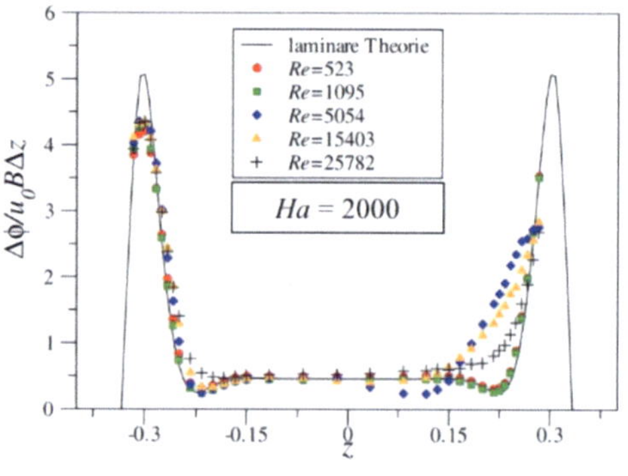

Fig. 23: Mean velocity profiles for Hunt flow at aspect ratio A=1/3.

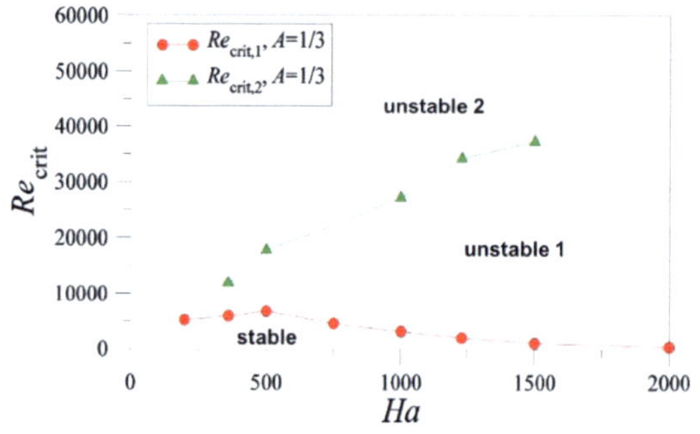

Fig. 24: Critical Reynolds number as a function of Ha for aspect ratio A=1/3.

References

[1] McCarthy, K. A.; Abdou, M. A.; "Analysis of liquid metal MHD flow in multiple adjacent ducts using an iterative method to solve the core flow equations", Fusion Engineering and Design, vol. 13, pp. 363-380, 1991.

[2] Molokov, S.; "Fully developed liquid-metal flow in multiple rectangular ducts in a strong uniform magnetic field", European Journal of Mechanics, B/Fluids, vol. 12, no. 6, pp. 769-787, 1993.

[3] Mistrangelo, C.; Bühler, L.; "Electric Flow Coupling in the HCLL Blanket Concept",

Fusion Engineering and Design, vol. 83, pp. 1232-1237, 2008.

[4] Mistrangelo, C.; Bühler, L.; "Electromagnetic flow coupling for liquid metal blanket applications", *Fusion Engineering and Design,* Vols. 109-111, pp. 1452-1457, 2016. doi: 10.1016/j.fusengdes.2015.11.052

[5] Mistrangelo, C.; Bühler, L.; "MHD phenomena related to electromagnetic flow coupling", *Magnetohydrodynamics,* in print, 2017.

[6] Mistrangelo, C.; Bühler, L.; "Overview of MHD phenomena related to electro-magnetic flow coupling", in Proceedings of the 10th PAMIR International Conference Fundamental and Applied MHD, Cagliari - Italy, June 20 - 24, 2016.

[7] Bühler, L.; Mistrangelo, C.; Electrically coupled 3D MHD flows in manifolds for helium cooled fusion blankets, 2016.

[8] Aubert, J.; Aiello, G.; Jonquares, N.; Puma, A.L.; Morin, A.; Rampal, G.; "Development of the water cooled lithium lead blanket for DEMO," *Fusion Engineering and Design,* vol. 89, no. 7-8, pp. 1386-1391, 2014.

[9] Mistrangelo, C.; Bühler, L.; "Development of a Numerical Tool to Simulate Magnetohydrodynamic Interactions of Liquid Metals with Strong Applied Magnetic Fields," *Fusion Science and Technology,* vol. 60, no. 2, pp. 798-803, 2011.

[10] Bühler, L.; Mistrangelo, C.; "MHD flow and heat transfer in model geometries for WCLL blankets," in Proceedings of the 29th Symposium on Fusion Technology SOFT, Prague, Czech Republic, Sept. 5-9, 2016.

[11] Mistrangelo, C.; Bühler, L.; "Magnetoconvective flows in rectangular ducts with internal cylindrical obstacles", in *17th MHD-Days 2016, Göttingen, November 30 – December 2,* 2016.

[12] Mistrangelo, C.; Bühler, L.; "Magnetoconvective instabilities in horizontal cavities", *Physics of Fluids,* vol. 28, no. 2, pp. 024104 1-15, 2016.

[13] Bühler, L.; Mistrangelo, C.; "MHD flow and heat transfer in model geometries for WCLL blankets", *Fusion Engineering and Design,* vol. in print, 2017.

[14] Bühler, L.; Mistrangelo, C.; "Theoretical studies of MHD flows in support to HCLL design", *Fusion Engineering and Design,* Vols. 109-111, pp. 1609-1613, 2016.

[15] Mistrangelo, C.; Bühler, L.; Koehly, C.; "Influence of modifications of HCLL blanket design on MHD pressure losses", *Proceedings of the 29th Symposium on Fusion Technology SOFT, Prague, Czech Republic, Sept. 5-9,* 2016.

[16] Hunt, J. C. R.; "Magnetohydrodynamic flow in rectangular ducts", *Journal of Fluid Mechanics,* vol. 21, pp. 577-590, 1965.

[17] Chowdhury, V.; Bühler, L.; Mistrangelo, C.; "Experimental investigations of instabilities in Hunt-type MHD flows", *Proceedings of the 10th PAMIR International Conference Fundamental and Applied MHD, Cagliari - Italy, June 20 - 24,* 2016, pp. 145-149.

[18] Chowdhury, V.; "Experimentelle Untersuchungen von magnetisch induzierten Instabilitäten in MHD Kanalströmungen", Dissertation, Karlsruhe Institute of Technology, 2016.

[19] Priede, J.; Arlt, T.; Bühler, L.; "Linear stability of magnetohydrodynamic flow in a square duct with thin conducting walls", *Journal of Fluid Mechanics,* vol. 788, pp. 129-146, 2016. doi:10.1017/jfm.2015.709

[20] Arlt, T.; Priede, J.; Bühler, L.; "Linear stability of magnetohydrodynamic flow in a square duct with thin conducting walls", *Proceedings of the 10th PAMIR International Conference Fundamental and Applied MHD,*

Cagliari - Italy, June 20 - 24, 2016, pp. 155-159.

[21] Arlt, T.; Priede, J.; Bühler, L.; "The effect of finite-conductivity Hartmann walls on the linear stability of Hunt's flow", *Journal of Fluid Mechanics,* submitted 2017.

[22] Arlt, T.; Priede, J.; Bühler, L.; "Influence of thin finitely conducting walls on the linear stability of magnetohydrodynamic flows", *Magnetohydrodynamics,* in print 2017.

[23] Krasnov, D.; Boeck, T.; Braiden, L.; Molokov, S.; Bühler, L.; "Numerical simulations of MHD flow transition in ducts with conducting Hartmann walls : Limtech Project A3 D4 (TUI)", KIT Scientific Reports 7713, KIT Scientific Publishing, Karlsruhe, 2016. doi:10.5445/KSP/1000057744

[24] Braiden, L.; Krasnov, D.; Molokov, S.; Boeck, T.; Bühler, L.; "Transition to turbulence in Hunt's flow in a moderate magnetic field", *Europhysics Letters,* vol. 115, no. 4, p. 44002, 2016.

Read more:

[25] Arlt, T.; Priede, J.; Bühler, L. (2016). Numerical investigation of magnetically induced instabilities. 17th MHD Days, Göttingen, November 30 - December 2, 2016.

[26] Braiden, L.; Krasnov, D.; Molokov, S.; Boeck, T.; Bühler, L. (2016). Transition to turbulence in Hunt's flow. Proceedings of the 10th PAMIR International Conference on Fundamental and Applied MHD, Cagliari, I, June 20-24, 2016, 27-31, University of Cagliari.

[27] Bühler, L.; Mistrangelo, C. (2016). Electrically coupled 3D MHD flows in manifolds for helium cooled lead lithium fusion blankets. 17th MHD Days, Göttingen, November 30 - December 2, 2016.

[28] Ehrhard, S. (2016). Numerische Simulation von MHD-Strömungen in inhomogenen und instationären Magnetfeldern (KIT Scientific Reports ; 7718). Dissertation. KIT Scientific Publishing, Karlsruhe. doi:10.5445/KSP/1000055453

[29] Koehly, C.; Bühler, L. (2016). Fabrication issues of sandwich-like flow channel inserts for circular pipes. 22nd Topical Meeting on the Technology of Fusion Energy (TOFE 2016), Philadelphia, PA, August 22-25, 2016.

A Benchmark on Dose Rate Calculations for PWR Spent Fuel Assembly

Barbara Vezzoni, Fabrizio Gabrielli, Andrei Rineiski

Introduction

In November 2014, the U.S. Department of Energy (DOE) and the French Commissariat à l'Énergie Atomique et aux Énergies Alternatives (CEA) proposed to the NEA Expert Group on Advanced Fuel Cycle Scenarios (WPFC/AFCS) a benchmark on dose rate calculations for Irradiated Fuel Assembly. CEA and DOE conducted a comparative study on dose rate calculations for typical Pressurized Water Reactor (PWR) spent fuel assemblies (with UOX and MOX fuels). The goal was to verify each organisation's gamma dose rate calculation methodology, especially for cases in which quantitative measurements of proliferation resistance are desired. The 30-year dose rates calculated by CEA and DOE were roughly three times lower than that of the reference study [1, 2]. In order to verify those results the benchmark has been extended to more international participants in the framework of NEA collaborations. Two phases of the study have been considered:

1) A verification phase – concluded on February 2017 – focusing on code-to-code comparisons for the dose rate calculated after 30 y and 3.7 y decay of UOX and MOX fuel [3]; and

2) A validation phase – on going – oriented to the comparison against experimental data available in literature [4-6].

The KIT/IKET (TRANS group) has contributed to the verification phase of the benchmark by assessing a dedicated dose rate calculation methodology using stochastic and deterministic codes. Related to this activity, the TRANS group has supported a master thesis work (C. Corrà) carried out in collaboration with Politecnico di Milano (Italy) in the frame of the GENTLE (Graduate and Executive Nuclear Training and Lifelong Education) project of the EU [7].

Background

The plutonium that is created in UOX fuel during irradiation and sometimes recycled in the form of mixed oxide (MOX) fuel can be an acquisition target for those wishing to obtain it for weapons use. If the UOX and MOX fuel assemblies were recently discharged after normal residence times from civilian reactors, then the significant amount of gamma radiation from the fuel assemblies make them unattractive to theft, thereby providing an inherent barrier of self-protection. The U.S. Nuclear Regulatory Commission (NRC) and International Atomic Energy Agency (IAEA) consider the "self-protecting" dose rate to be 1 Sievert per hour (Sv/h) at 1 meter from the fuel assembly [8].

The dose rate from a spent fuel assembly decreases with time after discharge due to radioactive decay of the gamma-emitting isotopes, therefore, an accurate prediction of this dose rate after decades of cooling becomes more and more important for SNF remaining for longer time in interim storage as it is the case for several industrialized countries (U.S., France, Germany, ...).

The accurate prediction of this dose rate depends on factors such as the assembly's power history, composition, and geometry as well as the calculated gamma source and radiation deposited on the target. Therefore, in addition to gamma transport calculations, the depletion, decay, and gamma source calculation approaches need to be precisely carried out.

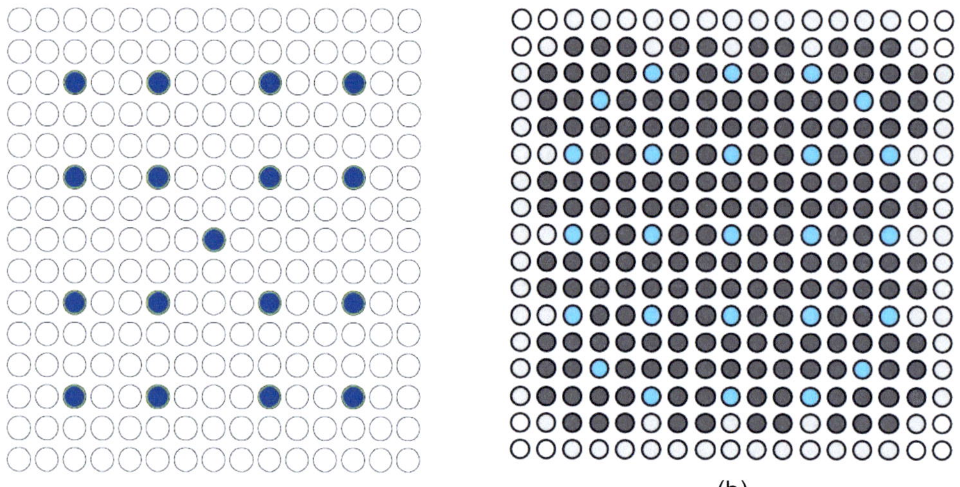

Fig. 1: a) 15x15 UOX fuel assembly (FA) with 17 water tubes, b) 17x17 MOX assembly, 24 tubes

Several international institutes have participated to the verification phase of the benchmark exercise: CEA from France, ANL from U.S.A., CIEMAT from Spain, AREVA from France, ENEA from Italy, CNL from Canada; INL from USA, VTT from Finland, SCK-CEN from Belgium and KIT from Germany. This large participation has allowed extensively comparing codes and methods for the three benchmark steps: depletion, decay and irradiation.

The TRANS group contribution is considered as part of the activities on-going at KIT concerning safety of Spent Nuclear Fuel (SNF) and waste characterization including inventory calculations and interim storage characterization.

Specifications of the verification benchmark

For the verification part of the benchmark, two spent fuel assembly models were analyzed: a 15x15 PWR assembly with UOX fuel (3.11 wt% ^{235}U and 33 GWd/tHM burnup) and a 17x17 PWR assembly with MOX fuel (average 9.6 wt% Pu and 60 GWd/tHM burnup). For each fuel assembly type, the depletion, the decay, and the radiation calculations have been performed. Fig. 1 shows the two assembly models.

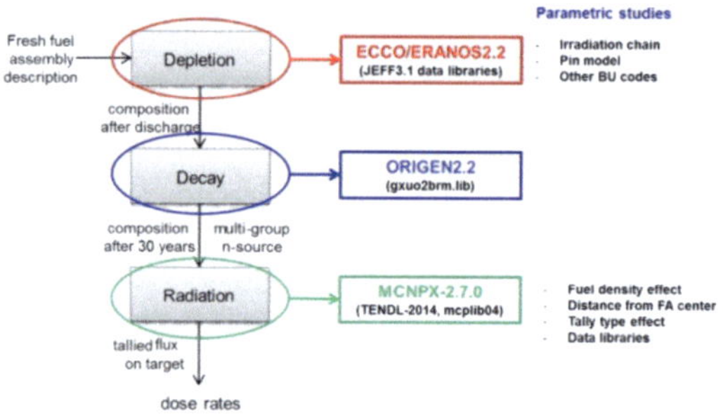

Fig. 2: KIT calculation scheme, codes and parametric studies considered

KIT methodology for dose rate calculations

In Fig. 2 the general calculation flow chart used at KIT for Dose Rate analyses is shown as well as a synthesis of the parametric studies carried out for the depletion and the radiation phases.

<u>Depletion</u>

For the Depletion part, the ECCO/ERANOS codes were used [9-10]. These codes have been developed for fast reactor studies but have been recently used also in Light Water Reactor (LWR) analyses by the TRANS group (e.g. on fuel cycle [11]). In order to take into account the self-shielding variation during LWR irradiation, at every burnup step, updated self-shielded microscopic cross section have been calculated by means of the ECCO cell code. Differences in the methodology adopted for the UOX and MOX fuels are reported in Table 1.

Several parametric studies have been carried out for this phase:

1) impact of burnup step during the ECCO/ERANOS calculation (34 x 20.2 → 68 x 10.1)

2) impact of modified pin model to take into account the rim effect (fuel zone subdivided in 5 concentric equal volume zones)

Table 1: KIT calculation scheme for depletion

Fuel type	Calculation scheme for depletion	Description
UOX	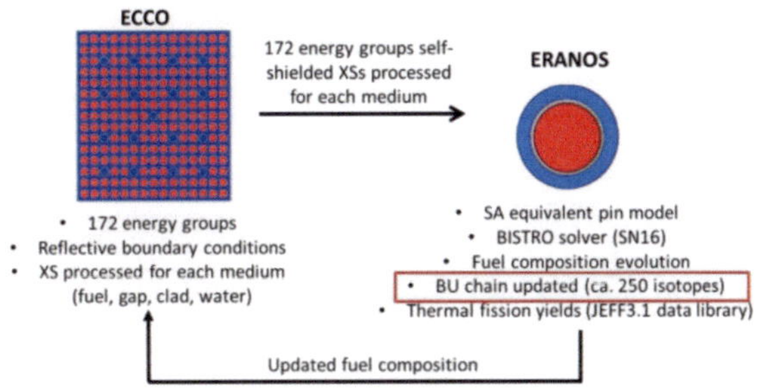	Microscopic cross-sections (XSs) calculated by ECCO actual subassembly (SA) geometry. UOX fresh fuel pins have the same enrichment. Therefore, the fuel assembly has been approximated with a SA equivalent pin ERANOS model for depletion. The 172 energy groups self-shielded XSs have been processed at every BU step (ca. 1.5 GWd/ton iHM each).
MOX	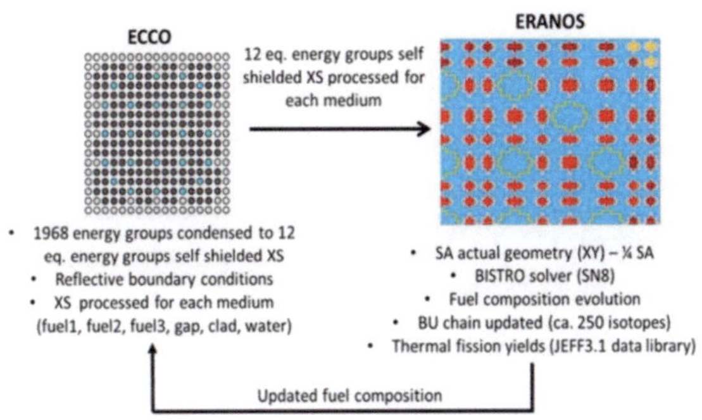	Microscopic cross-sections calculated by ECCO actual subassembly (SA) geometry. MOX fresh fuel pins do not have the same enrichment. Therefore, actual geometry has been considered for the ERANOS model for depletion. ¼ SA has been modeled using XY coordinates. The 12 equivalent energy groups effective XSs (condensed from 1968 gr.) for each burnable zone have been reprocessed at every BU step (ca. 1.5 GWd/ton iHM each).

3) use of TRAIN code (burnup KIT code [12]) taking into account decay heat contribution in the flux normalization

4) use of TRAIN code (burnup KIT code [12]) with different normalization schemes: a) at t=0, b) at the beginning of each sub-interval, c) at the beginning of each sub-interval but linear between each sub-interval.

The results have allowed improving the depletion procedure for LWR used in the group. An extended BU chain for ERANOS has been indeed implemented and tested [7].

Decay

The composition after depletion calculated with ECCO/ERANOS procedure (Table 1) is used as input (after rescaling to the fuel assembly total mass) for the ORIGEN2.2 code. The composition evolution during natural decay has then been modeled by ORIGEN as well as the photon energy emission rate in 18 energy groups. The composition and the emission rate after decay have been then provided as input for the irradiation step. An example of calculated gamma source after 30 y decay for the UOX case is shown in Fig. 3.

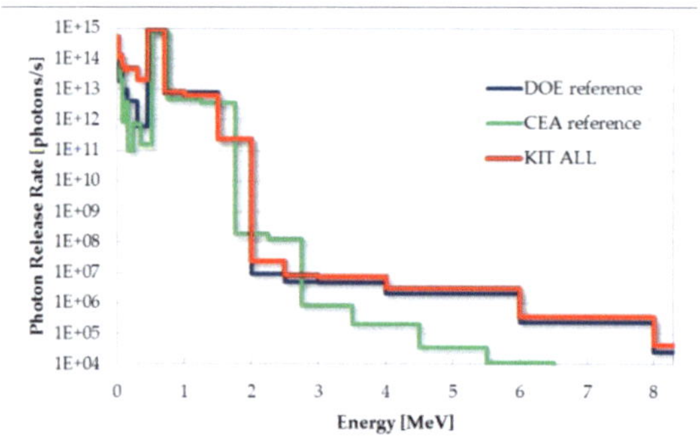

Fig. 3: UOX computed photon release rate in comparison with CEA and DOE reference plot. The different energetic structure of the CEA can be easily noticed on the green line trend.

composition and gamma source calculated with ORIGEN2.2 have been spread uniformly axially and radially in all the fuel pins. The assembly is located in air and the gamma dose is evaluated at 1m from the assembly faces at the axial mid plane (Fig. 4-b).

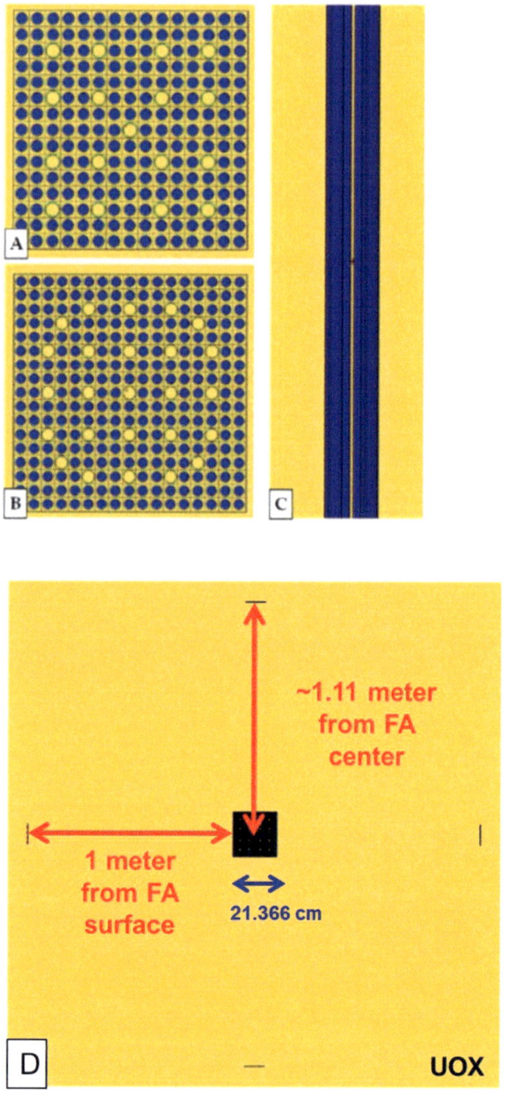

Fig. 4: A) UOX MCNPX mode (XY plane), B) MOX MCNPX mode (XY plane), C) UOX MCNPX mode (XZ plane), and D) UOX MCNPX mode (XY plane) plus tally locations.

Irradiation

3D heterogeneous fuel assembly models for UOX and MOX fuel have been assessed by using MCNPX-2.7.0 code (Fig. 4-A/B). The fuel

The reference KIT calculation has been done using TENDL14 library, Tally F2 (surface) averaged over 4 x 100 cm^2 surfaces located at the axial mid plane and 1m from the FA surface as in Fig. 4.C. In order to investigate the impact of different options, several parametric studies have been carried out also for this phase:

1) Impact of tally surface (Tally F2 – 400/100 cm^2)
2) Impact of tally type (Tally F4 vs. F2, namely volumetric vs. superficial)
3) Impact of TENDL distributions (TENDL-2012 vs. TENDL 2014)
4) Impact of fuel density
5) Impact of tally location (distances from the FA center, axial locations, etc.).

The results have allowed improving the gamma irradiation calculation for LWR used in the group. The same approach can be applied to other problems, e.g. dose rate calculation for CASTOR at interim storage.

Conclusions and outlook

The KIT/IKET (TRANS group) has contributed to the verification phase of the benchmark on dose rate calculations for Irradiated Fuel Assemblies carried out in the framework of the NEA/AFCS activities. The results obtained for the verification phase by all the participants are in very good agreement (std. dev. of 14% for UOX and 10% for MOX at 30 years). KIT results differ to the average value of 5% for the UOX case and 3% for the MOX case, respectively.

An overall summary of the results obtained during the verification phase will be published by the end of 2017 [3] including an analysis of the different sensitivity studies carried out to confirm this good agreement: homogeneous/heterogeneous assembly geometry, axial burnup distribution, pin-by-pin gamma source distribution, gamma source group structure, Bremsstrahlung effect, X radiation from beta sources, neutron dose rate contribution, but also tally geometry description, Monte Carlo flux estimators, flux to dose conversion factors, etc.

The NEA/AFCS activity is planned to go on with the second phase of the benchmark aiming at validating the calculation procedures, by a comparison with available experimental data of measured dose rates from commercial spent fuel through air after few years of cooling [4-6].

References

[1] Feng, B.; Hill, R.; Girieud, R.; Eschbach, R.; "Comparison of gamma Dose Rate calculations for PWR Spent Fuel assemblies", PHYSOR 2014, Kyoto, Japan, September 28 – October 3, 2014.

[2] Lloyd, W.R.; Sheaffer, M.K.; Sutcliffe, W.G.; "Dose Rate Estimates from Irradiated Light-Water Reactor Fuel Assemblies in Air", Lawrence Livermore National Laboratory, Livermore, CA, UCRL-ID-115199 (1994).

[3] Eschbach, R.; Feng, B.; Vezzoni, B.; Alvarez-Velarde, F.; Carliere, B.; Légere, V.; Rocchif, F.; Edwardsg, G.; Dixonh, B.; Pénéliaua, Y.; Girieuda, R.; Häkkineni, S.; Viitaneni, T.; Malambuj, E. M.; Gabriellic, F. and Cornetk S. et. al., "A Benchmark on Dose Rate Calculations for PWR Spent Fuel Assembly", Submitted to Global 2017.

[4] Willingham, C.E.; "Radiation Dose Rates from Commercial PWR and BWR Spent Fuel Elements," PNL-3954, Pacific Northwest Laboratory, October (1981). URL: https://inis.iaea.org/search/search.aspx?orig_q=RN:12642536

[5] Davis, R. B. 1980a. Data Report for the Nondestructive Examination of Turkey Point Spent Fuel Assemblies B02, B03, B17, B41 and B43. HEDL-TME-79-68. Hanford Engineering Development Laboratory, Richland,

Washington. URL: http://www.osti.gov/scitech/servlets/purl/5445903

[6] Davis, R. B. 1980b. Pre-Test Nondestructive Examination Data Summary Report on Turkey Point Spent Fuel Assemblies 001, 004 and 006 for the Climax-Spent Fuel Test, HEDL-TME-80-83. Hanford Engineering Development Laboratory, Richland, Washington. URL: http://www.osti.gov/scitech/servlets/purl/971850

[7] Corrà, C.; "COUPLED N-γ CALCULATIONS FOR FUEL CYCLE AND SAFETY STUDIES", Politecnico di Milano, Italy Master Thesis discussed on April 2016.

[8] U.S. Department of Energy, "Non-proliferation and Arms Control Assessment of Weapons-Usable Fissile Material Storage and Excess Plutonium Disposition Alternatives", DOE/NN-007 (1997).

[9] Rimpault, G., et al., "The ERANOS Code and Data System for Fast Reactor Neutronic Analyses", Proc. Int. Conf. PHYSOR 2002, Seoul, Korea, 7-10 October 2002.

[10] Rimpault, G., "Algorithmic features of the ECCO cell code for treating heterogeneous fast reactor assemblies", Proc. of Int. Conf. M&C, Portland, Oregon, 1-5 May, 1995.

[11] OECD/NEA, "The Effects of the Uncertainty of Input Parameters on Nuclear Fuel Cycle Scenario Studies", report under publication.

[12] RINEISKI, A., "Decay heat production in a TRU burner", Progress in Nuclear Energy 50 (2008) 377–381.

Read more:

[13] Wang, S.; Rineiski, A.; Guo, L.; "A New Lumped Thermal Resistance Heat Transfer Model for Fuel Pin Structure", Journal Nuclear Technology, Vol. 196, Nr. 3, pp. 588-597, December 2016. doi:10.13182/NT16-5

[14] Gabrielli, F.; Flad, M.; Gianfelici, S.; Li, R.; Krinventsev, V.; Maschek, W.; Matzerath Boccaccini, C.; Vezzoni, B.; Rineiski, A.; "Probabilistic Evaluation of the Post Dissassembly Energetics of a Hypothetical Core Disruptive Accident in a Sodium-Cooled SMR by using a Phenomenological Relationship Diagram", INES-5: 5th Int. Symposium on Innovative Nuclear Energy Systems, Tokyo, Japan, October 31 – November 2, 2016.

[15] Chen, X.-N.; Li, R.; Gabrielli, F.; Rineiski, A.; Andriolo, L.; Guo, L.; Castelliti, D.; Bubelis, E.; Bandini, G.; Sarotto, M.; "Recent MYRRHA Safety Studies with the SIMMER Code", NUTHOS-11: 11th Int. Topical Meeting on Nuclear Reactor Thermal Hydraulics, Operation and Safety, Gyeongju, Korea, October 9-13, 2016.

[16] Guo, L.; Chen, X.-N.; Rineiski, A.; "Further Development of COUPLE Code Taking Account of Gas-Liquid Interfacial Motion", NUTHOS-11: 11th Int. Topical Meeting on Nuclear Reactor Thermal Hydraulics, Operation and Safety, Gyeongju, Korea, October 9-13, 2016. Proceedings on USB-Stick, Paper No.N11P0072.

[17] Wang, S.; Massone, M.; Rineiski, A.; Merle-Lucotte. (2016). Analytical investigation of the draining system for a molten salt fast reactor. 11th International Topical Meeting on Nuclear Thermal-Hydraulics, Operation and Safety (NUTHOS-11), Gyeongju, Korea, October 9-13, 2016. Proceedings on USB-Stick, Paper No. N11A0341.

[18] Chen, X.-N.; "Hydrodynamic Analogy to Special Relativity", World Journal of Mechanics, Vol. 6, pp. 406-418, October 2016. doi:10.4236/wjm.2016.610029

[19] Palmiotti, G.; Salvatores, M.; Hursin, M.; Kodeli, I.; Gabrielli, F.; Hummel, A.; "New approaches to provide feedback from nuclear and covariance data adjustment for effective improvement of evaluated nuclear data files Int. Conf. on Nuclear Data for Science and

Technology (ND2016), Bruges, Belgium, September 11-16, 2016.

[20] Fei, T.; Vezzoni, B.; Marchetti, M.; van Rooijen, W. F. G.; Zhang, Y. (2016). Neutronics benchmark for EBR-II shutdown heat removal test SHRT-45R. Proceedings of PHYSOR 2016 : Unifying Theory and Experiments in the 21st Century, Sun Valley, Idaho, May 1-5, 2016, 221-230, American Nuclear Society, LaGrange Park (IL).

[21] Vezzoni, B.; Marchetti, M.; Gabrielli, F.; Rineiski, A. (2016). IAEA EBR-II neutronics benchmark: Impact of modeling options on KIT results. Proceedings of PHYSOR 2016 : Unifying Theory and Experiments in the 21st Century, Sun Valley, Idaho, May 1-5, 2016, 719-728, American Nuclear Society, LaGrange Park (IL).

[22] Gabrielli, F.; Rineiski, A.; Talamo, A.; Gohar, Y.; Pyeon, C. H. (2016). Deterministic analyses of the phase-I-kinetic experiments in the KUCA subcritical A-core configurations. Proceedings of PHYSOR 2016 : Unifying Theory and Experiments in the 21st Century, Sun Valley, Idaho, May 1-5, 2016, 1373-1382, American Nuclear Society, LaGrange Park (IL).

[23] Talamo, A.; Gohar, Y.; Gabrielli, F.; Rineiski, A.; Pyeon, C. H. (2016). Coupling Sjöstrand and Feynman methods in prompt neutron decay constant analyses. Proceedings of PHYSOR 2016 : Unifying Theory and Experiments in the 21st Century, Sun Valley, Idaho, May 1-5, 2016, 2253-2263, American Nuclear Society, LaGrange Park (IL).

[24] Gabrielli, F.; Flad, M.; Gianfelici, S.; Li, R.; Krivensev, V.; Maschek, W.; Matzerath Boccaccini, C.; Vezzoni, B.; Rineiski, A. (2016). Application of a probabilistic relationsship diagram for PDE mechanical energy release evaluation fter HCDA in a sodium-cooled SMR. Proceedings of PHYSOR 2016 : Unifying Theory and Experiments in the 21st Century, Sun Valley, Idaho, May 1-5, 2016, 211-220, American Nuclear Society, LaGrange Park (IL).

[25] Vezzoni, B.; Marchetti, M.; Andriolo, L.; Gabrielli, F.; Chen, X.-N.; Matzerath Boccaccini, C.; Rineiski, A.; Maschek, W.; Morita, K.; Arima, T.; "SIMMER/PARTISN Analyses of EBR-II Shutdown Heat Removal Tests", Proc. of Int. Conf. PHYSOR 2016, Sun Valley, ID, USA, May 1–5, 2016.

[26] Guo, L.; Li, R.; Wang, S.; Flad, M.; Maschek, W.; Rineiski, A. (2016). Numerical investigation of SIMMER code for fuel-coolant interaction. International Journal of Hydrogen Energy, 41 (17), 7227-7232. doi:10.1016/j.ijhydene.2016.01.080

[27] Chen, X.-N.; Rineiski, A. (2016). Asymptotic radial fuel shuffling mode for accelerator driven transmuter. International Journal of Hydrogen Energy, 41 (17), 7112-7118. doi:10.1016/j.ijhydene.2016.01.126

[28] Li, R.; Maschek, W.; Matzerath Boccaccini, C.; Vezzoni, B.; Flad, M.; Rineiski, A. (2016). Impact of the Bell-Plesset instability on centralized sloshing in pool geometry. International journal of hydrogen energy, 41 (17), 7126-7131. doi:10.1016/j.ijhydene.2016.01.152

[29] Rohde, M.; Peeters, J. W. R.; Pucciarelli, A.; Kiss, A.; Rao, Y. F.; Onder, E. N.; Muehlbauer, P.; Batta, A.; Hartig, M.; Chatoorgoon, V.; Thiele, R.; Chang, D.; Tavoularis, S.; Novog, D.; McClure, D.; Gradecka, M.; Takase, K.; "A Blind, Numerical Benchmark Study on Supercritical Water Heat Transfer Experiments in a 7-Rod Bundle", Journal of Nuclear Engineering and Radiation Science, Vol. 2, April 2016.

[30] Chen, X.-N.; Rineiski, A.; Gabrielli, F.; Andriolo, L.; Vezzoni, B.; Li, R.; Maschek, W.; Kiefhaber, E. (2016). Fuel-steel mixing and radial mesh effects in power excursion simulations. Annals of Nuclear Energy, 90, 26-31. doi:10.1016/j.anucene.2015.11.041

[31] Li, R.; Maschek, W.; Matzerath-Boccaccini, C.; Marchetti, M.; Kriventsev, V.; Rineiski, A. (2016). Bell-Plesset instability analysis for an inward centralized sloshing. Nuclear Engineering and Design, 297, 312-319. doi: 10.1016/j.nucengdes.2015.12.010

[32] Kriventsev, V. (2016). A rule of minimal energy dissipation in modeling of channel flow turbulence. Numerical Heat Transfer, Part B: Fundamentals, 69 (2), 85-95. doi: 10.1080/10407790.2015.1093793

[33] Hartig, M.; Schulenberg, T.; Batta, A.; "Numerical Study of Fluid-Structure Interaction with Heat Transfer at Supercritical Pressure in a Fuel Rod Assembly", Journal of Nuclear Engineering and Radiation Science, Vol. 2, pp. 1-12, January 2016.

[34] Tiphine, M.; Vezzoni, B.; Alvarez-Velarde, F.; Carlier, B.; Wojtaszek, D.; Glinatsis, G.; Brolly, A.; Dixon, B.; Eschbach, R.; Coquelet-Pascal, C.; Freynet, D.; Edwards, G.; Hyland, B.; Häkkinen, S.; Viitanen, T.; Gabrielli, F.; Ono, K. and Cornet, S.; "The Effects of the Uncertainty of Input Parameters on Nuclear Fuel Cycle Scenario Studies", Proc. Int. Conf. 14th IEMPT (NEA-Exchange Meeting on P&T), San Diego, USA, 17-20 October 2016.

[35] Schwenk-Ferrero, A.; "Key Indicator Scales and Comment on Utility/Value Function Selection" ,Annex II : The Mathematical Background and Details of MCDA Methods" in APPLICATION OF MULTI-CRITERIA DECISION ANALYSIS METHODS TO COMPARATIVE EVALUATION OF NUCLEAR ENERGY SYSTEM OPTIONS, IAEA Nuclear Energy Series

[36] Schwenk-Ferrero, A.; "OECD-NEA scenarios: Evaluation and aggregation judgement measures for performance comparison of advanced nuclear fuel cycles", in APPLICATION OF MULTI-CRITERIA DECISION ANALYSIS METHODS TO COMPARATIVE EVALUATION OF NUCLEAR ENERGY SYSTEM OPTIONS Chapter 5.8 Development scenarios, IAEA Nuclear Energy Series

[37] Schulenberg, T. (2016). Alternativen zur nuklearen Entsorgung. Landesfachausschuss Forschung, Technologie und Energie, Frankfurt, 15.Oktober 2016.

[38] Schwenk-Ferrero, A. (2016). Evaluation and aggregation judgement measures for performance comparison of advanced nuclear fuel cycles. International Atomic Energy Agency, Wien, A, June 16, 2016.

[39] Vezzoni, B.; Marchetti, M.; Andriolo, L.; Gabrielli, F.; Chen, X. N.; Matzerath Boccaccini, C.; Rineiski, A.; Maschek, W.; Morita, K.; Arima, T. (2016). SIMMER/PARTISN analyses of EBR-II shutdown heat removal tests. Proceedings of PHYSOR 2016 : Unifying Theory and Experiments in the 21st Century, Sun Valley, Idaho, May 1-5, 2016, 3973-3982, American Nuclear Society, LaGrange Park (IL).

Group: Accident Analysis

Analysis of Design Basis and Severe Accidents

Alexei Miassoedov, Giancarlo Albrecht, Thomas Cron, Philipp Dietrich, Beatrix Fluhrer, Jerzy Foit, Stephan Gabriel, Xiaoyang Gaus-Liu, Frank Kretzschmar, Markus Schwall, René Stängle, Mike Vervoortz, Thomas Wenz

Introduction

The analysis of severe accidents in LWRs at IKET-UNA is focused on the in- and ex-vessel core melt behavior. The overall objective is to investigate the core melt scenarios from the beginning of core degradation to melt formation and relocation in the vessel, possible melt dispersion to the reactor cavity and to the containment, and finally corium concrete interaction and corium coolability in the reactor cavity.

The experimental platform includes three experimental facilities:

- LIVE to investigate the melt pool behavior in the RPV lower head;
- DISCO to study the melt dispersion to the reactor cavity and direct containment heating;
- MOCKA to study molten corium concrete interaction.

The results of the experiments are being used for the development and validation of codes applied for safety assessment and planning of accident mitigation concepts, such as MELCOR and ASTEC. The strong coupling between the experiments and analytical activities contributes to a better understanding of the core melt sequences and thus improves safety of existing reactors by severe accident mitigation measures and by safety installations where required.

LIVE experiments

The main objective of the LIVE program is to study the late in-vessel core melt behavior and core debris coolability both experimentally in large scale 2D and 3D geometry and in supporting separate-effects tests.

Two large-scale test facilities are involved in the LIVE experimental programme, the hemispheric LIVE3D and the semi-circular LIVE2D, are both 1 m in diameter, the 1:5 length scale of a reactor vessel. The LIVE2D test vessel has a width of 12 cm. Its two vertical sidewalls are thermally insulated. The curved steel vessel walls of LIVE3D and LIVE2D have a thickness of 2.5 cm, and are enclosed in a cooling vessel. Simulant materials of core melt are water and nitrate salt mixture with either non-eutectic molar composition of 80% KNO_3-20% $NaNO_3$ or eutectic composition (50% KNO_3-50% $NaNO_3$). The solidification range of the non-eutectic mixture is 224 °C to 284 °C. To simulate different patterns of melt relocation, there is a melt preparation furnace and pouring arrangement. Resistant heating elements simulate the decay heat source. The power of each heating element can be individually adjusted to realize either volumetric heating in the whole pool or defined volume-related heating density. The curved sidewall can be cooled externally with water either in large flow rate or in nucleate boiling condition. Both top insulation and top cooling condition are available. The insulation lid simulates a closed hot environment in the reactor vessel, whereas the top cooling lid mimics the situation of water injection on the melt surface.

The measured data includes 3D melt temperature distribution, boundary layer temperatures and crust temperatures, 3D wall temperature distribution at the inner and outer surface, a precise temperature vertical profile at a certain radial position, cooling water parameter of the external and top cooling. In the case of test with insulation lid, the temperature in the gas gap between the melt surface and the insulation lid is measured. The melt sample before, during and at the end of a test can be taken for the determination of composition and the melting temperature. Crust profile at the end of an experiment can also be measured after the liquid melt is extracted from the vessel.

In 2016 upgrading of LIVE2D facility to study the thermos-hydraulic and heat transfer behavior of a stratified melt pool was performed. The results of the OECD MASCA project demonstrate a very complex process of layer partition between oxide and metal in the corium with the consequence that two or three melt layers may coexist: one heavy metallic layer at the bottom, an oxide layer in between and one light metallic layer at the top can occur during the evolution of a melt pool. The sidewall heat transfer in the light metallic layer atop of the oxide layer is considered especially vital for the coolability of the vessel, since this layer receives a large portion of the heat from the underlying oxide layer and has only limited heat transfer area on the sidewall. The vessel wall becomes non-coolable when the heat flux on the wall exceeds the cooling limit of the external coolant.

The demand on the upgrading of LIVE2D facility includes the selection of two safe and stable simulant materials which can separate from each other physically and chemically, the installation of a transparent plate on the vertical sidewall for the optic observation of the melt pool, additional instrumentation to capture the detailed character of two-layer heat transfer and at last the additional safeguard measures to guarantee personal safety and safe operation.

Selection of simulant materials

The eutectic nitrate salt with the molar composition of 50% $NaNO_3$ - 50% KNO_3 as the oxide melt and a thermal oil "Therm 240" as the upper metallic layer are selected. The two materials differ large in density and are not solvable in another. The long-term exposure test, labor-scale hot test and combustion test of the two-component mixture confirmed the stability of the two materials with each other and with vessel installations. The pre-calculation of the heat transfer behaviour in the two-layer melt in LIVE2D geometry with different upper layer thickness provided further information of the possible operational temperature ranges. The liquid-state temperature of the nitrate is 224°C-380°C and the thermal oil can be used from ambient temperature to 250°C.

Replacement of LIVE2D vertical wall with a glass wall

The vertical wall with transparent material can provide very important information of the melt during the transient process of layer separation, interlayer crust formation and stability, and the cooling effect of a gap along the vessel wall in the oxide layer. Also turbulent flow patterns of the liquid melt during the steady-state melt can be obtained. The transparent material should withstand thermal shock, large temperature gradient and be chemically stable with the simulant melt material. Borosilicate glass was firstly employed (Fig. 1). However, this material cannot withstand the dramatic temperature gradient projected from the melt pool. Fused quartz material with very small thermal expansion rate is a more suitable material and will be applied. Another challenge is to guarantee the tightness between the plate and the curved vessel wall in a varying melt temperature under the different thermal expansion rate of the materials. A double-layer construction with each layer of 15 mm-thick quartz plate and an interlayer gap of about 18 mm is designed. The plates are positioned in a flange with enough freedom of thermal expansion. A high-temperature silicon sealing is applied between the plates and the flange. With the double-layer construction the thermal loss through the plates is estimated less than 3% of the heat input.

Additional instrumentations

Following instrumentation was added to the facility:

- 3 mobile thermocouple trees mounted on the top lid to measure the temperature profile in the upper layer,
- 3 pairs of thermocouple on the vessel inner and outer surface for the determination of heat flux distribution in the upper layer,
- 9 thermocouples, which are vertically arranged near the symmetric axis, to measure the vertical temperature profile in the melt.

Safety measures

Two additional safety measures were implemented for the melt stratification test:

- Power supervision and automatic function in case of heating element failure or fire,
- Oil vapor extraction installations above the test vessel. Oil vapor can the collected and brought out of the experimental building.

Besides the upgrading of LIVE2D facility, a LIVE3D test was performed using water as simulant material. Both top insulation and top cooling conditions were realised. There were two pool heights during the top insulation condition. For each boundary condition, two levels of heat input were performed. The LIVE3D water test complemented the previous tests in LIVE program and other studies, and will provide data for the understanding of the heat transfer of crust-free melt.

Fig. 1: Upgraded LIVE2D facility with the transparent side wall.

MOCKA experiments

The MOCKA facility is designed to investigate the corium/concrete interaction in an anticipated core melt accident in LWRs, after the metal melt is layered beneath the oxide melt. The experimental focus is on the cavity formation in the basemat and the risk of a long-term basemat penetration by the metallic part of the melt.

In MOCKA experiments cylindrical concrete crucibles with an inner diameter of 25 cm are used. Both the sidewall and the bottom are instrumented with Type K thermocouple assemblies to determine the concrete erosion as a function of time. The initial melt consists of 42 kg Fe together with 4 kg Zr, overlaid by 68 kg oxide melt (initially 56 wt.% Al_2O_3, 44 wt.% CaO). The melt temperature at start of interaction is approximately 1900 °C. The CaO admixture lowers the solidus temperature and the viscosity of the oxide melt. The resulting solidus temperature of approx. 1360 °C is sufficiently low to prevent a formation of an initial crust at the oxide/concrete. To allow for a long-term interaction, internal heating is provided by alternating additions of thermite and Zr metal to the upper oxide layer of the stratified melt. The heat generated by the thermite reaction and the exothermal oxidation reaction of Zr is mainly deposited in the oxide. Due to density-driven phase segregation the metal melt at the bottom of the crucible is fed by the enthalpy of Fe melt which is generated in the oxide phase by the thermite reaction of the added thermite. Approximately 75 % of the heating power is deposited in the oxide phase and 25 % in the iron melt. In this way a rather prototypic heating of both melt phases is achieved. The heating method used in MOCKA allows investigation of MCCI process with reinforced concrete.

In 2016, two MOCKA experiments with reinforced concrete have been performed. An upside-down view of the rebar structure used in MOCKA experiments is given in Fig. 2. The first test, MOCKA 5.8, was a test with siliceous concrete and 12 wt% rebars and the second test, MOCKA 7.3 was a test with LCS (Limestone Common Sand) concrete and 12 wt% rebars. The findings of the two tests confirm the results of the previous performed experiments.

Fig. 2: Upside-down view of the rebar structure of MOCKA experiments.

The efficient heat transfer from the metal melt to the concrete leads to a fast decrease in temperature. The metal melt temperature near to the concrete surface decreases below the solidus temperature of the iron melt, which in turn give rise to formation of a crust. From the post-test analyses there are no indications that the formed crusts are permeable to gravels. Therefore, the crust will temporary prevent further melt front progression. In all experiments, a considerable delay of the arrival of the metal melt was observed by an electrical circuit with in concrete embedded electrodes if compared with the thermocouples findings, Fig. 3 (for MOCKA 5.6 test). Consequently, the estimated progression rate by the thermocouple readings during the early phase of the interaction does not represent the progression of the encrusted metal melt into the concrete. A considerable heat-up of a rather thick layer of concrete ahead the crust leads to a loss of structural stability of the concrete.

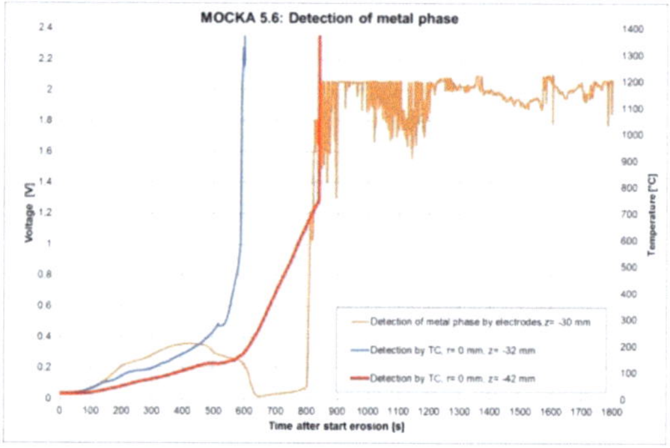

Fig. 3: Detection of the metal melt front.

The rebars in the concrete elevate the melt/concrete interface temperature up to the melting temperature of the reinforcing steel, i.e. 1528 °C. This should result in higher melt pool temperatures than during MCCI with concrete without reinforcement for which a concrete decomposition temperature of approximately 1300 °C was estimated. After a fast initial decrease in melt temperature, the temperature of the oxide melt clusters around 1600 °C in both experiments. As the thermal properties of the pure concrete and that of the reinforced concrete do not differ much and, in addition, the same internal heating power was generated, in all experiments under consideration, the former estimated decomposition temperature of the pure concrete cannot be used to prescribe the temperature boundary condition for the MCCI process. Based on MOCKA experimental findings the melt/concrete interface temperature which is relevant to the MCCI process must be as least as high as the melting temperature of the commonly used reinforcing steel, i.e. approximately 1528 °C.

MIT3BAR experiment in the MOCKA test facility

In the frame of the French MIT3BAR program one large-scale experiment has been performed in September 2016 in the MOCKA test facility in cooperation with CEA to investigate the technical feasibility and realization of a modified CometPC concept in existing power plants, especially with respect to the geometric limitations.

The original CometPC concept of corium cooling is developed to arrest and cool ex-vessel corium melts in the case of core melt accidents. After erosion of a sacrificial concrete layer the melt is passively flooded by a layer of porous, water filled concrete (CometPC = Comet Porous Concrete). The open porosity of the concrete is adjusted for the appropriate coolant water flow into the melt. Horizontal distribution of the water is achieved by a second, high porosity concrete layer underneath. The modification of the original COMETPC concept consists of higher thicknesses of the porous concrete layers.

In the experiment the cooling of an oxidic and metallic corium melt in a cylindrical cavity (inner diameter 920 mm) with walls made from castable refractory and the modified cooling device at the bottom was investigated (Fig. 4). The cooling device consisted of a round, flat concrete sump made of siliceous concrete with an above located high porosity concrete layer with a height of 8 cm and a low porosity concrete layer with a height of 23 cm. This layer with low porosity limited the water flow to the melt to a certain value. The cooling device was completed by a 5 cm standard siliceous concrete layer on top of the low porosity concrete layer.

The melt was generated by a thermite reaction of 1150 kg thermite/CaO mixture. The resulting melt consisted of 430 kg iron melt, overlaid by 720 kg oxide melt. For the first time for a CometPC-type experiment, both oxidic and metallic melt layers were heated by alternating additions of thermite and Zr metal. The mass ratio of the added thermite and Zr as well as the time sequence of the additions were estimated to provide a prototypic heating of both melt phases.

Fig. 4 MIT3BAR test crucible

After a rather homogeneous erosion of the sacrificial concrete layer the melt was passively flooded by bottom injection of coolant water through two porous concrete layers with different heights and porosities. The selected effective water pressure of 0.2 bar over the static pressure of the melt at the time of melt/ water contact provided a water inflow rate of approximately 1.4 l/s. The resulting evaporation process created a porous structure of the melt, from which the heat was easily extracted (Fig. 5). The sufficiently high porosity of the solidified oxide melt provided a long-time coolable configuration of the oxide melt. The experiment demonstrated that the modified CometPC concept is able to stop and cool the melt.

Fig. 5: Ejection of melt during the cooling process

DISCO experiments

KIT in cooperation with IRSN conducted a series of experiments in the DISCO facility, with geometry roughly typical of a PWR at reduced scale, involving ex-vessel Fuel Coelant Interaction (FCI) with melt injection from a pressurized vessel (~10 bars). The melt was a mixture of Fe and Al_2O_3 (10-16 kg). One experiment was conducted in the LACOMECO frame. Two tests led to spontaneous explosions, the strengths of which were moderate due to late triggering of the explosion after the start of the ejection. The calculations performed with MC3D indicate two possible triggering events: a possible flow back of the water on the jet and the change of configuration of melt flow at the break (Fig. 6). Another outcome is the confirmation of the strong oxidation of the iron and hydrogen production, similarly to DCH tests (where there was no water in the pit). A visualization of melt-water interaction is needed to compare it with the calculation.

To allow visualization, the DISCO facility was upgraded by a vessel with glass windows is inserted instead of a steel pit (Fig. 7). The water pool has a cross-section of 0.7 m x 0.7 m and a height of 1.40 m. The maximum volume is 0.686 m³. However, in order to observe the water level rise, the water volume should not be more than 0.63 m³, preferably about 0.52 m³ or less. In comparison, the water reservoir in the cylindrical pit of the previous FCI experiments has a diameter of 0.54 m and a height of 0.54 m corresponding to a volume of 0.124 m³.

Three sides of the vessel are provided with glass windows the fourth side is made of steel and will be used for instrumentation, e.g. pressure transducers and thermocouples at different positions and a level meter to measure the water movement. Two high-speed cameras and a video camera observe and record the interaction. After the experiment a post-test debris characterization (fragmentation or cake, size distribution) will be performed.

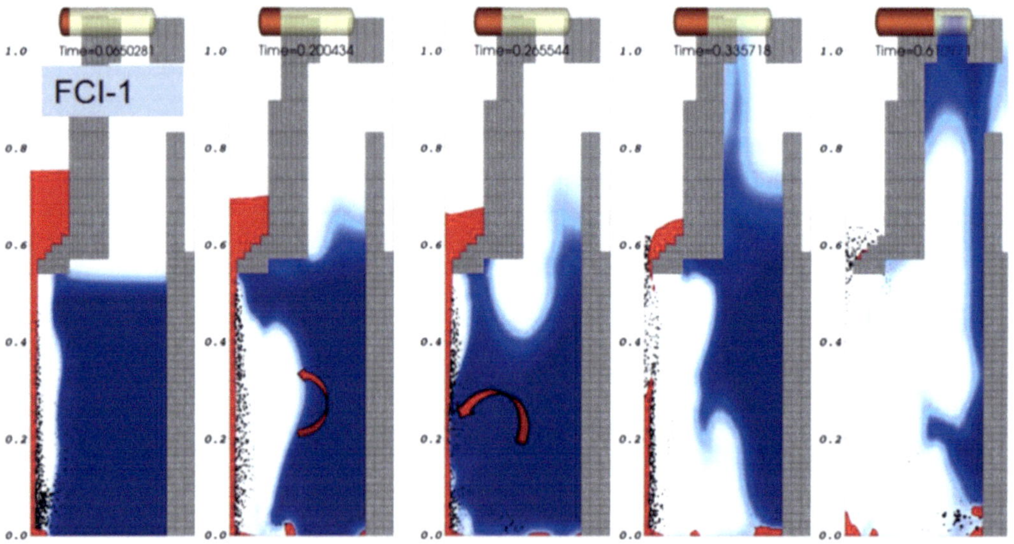

Fig. 6:. MC3D-Simulation of the DISCO FCI-1 experiment

Fig. 7: DISCO vessel for visualization of the melt jet interaction with water

To concentrate on the visualization some simplifications are necessary. A steam atmosphere would cause fogging on the windows, thus affecting the view through the glass. Therefore, an air atmosphere is required. The pressure in the containment can be as high as in the FCI tests: 2 bar abs. As driving gas in the Reactor Cooling System (RCS) nitrogen instead of steam would be preferable, because no preheating of the RCS would be necessary. Table 1 shows the conditions of the experiment which are similar to these of the LACOMECO test. The DISCO-Q experiment is in preparation and will be carried out in April 2017.

Application of the MELCOR code

One of tasks of MELCOR code application was to compare the results of analysis of an accident sequence in a generic German PWR between MELCOR and ATHLET-CD in cooperation with HZDR. A SBLOCA accident sequence without secondary and with HPIS was selected for this comparison.

IKET-UNA created the MELCOR input on the basis of the provided documentation by HZDR. The MELCOR model is a two loop model, which is similar to the two loop ATHLET-CD model: a single loop, which represents the loop connected to the pressurizer and a triple loop, which represents the remaining three loops of the plant. Fig. 8 shows the nodalization scheme of the model. All main components and plant systems necessary for the simulation of the SBLOCA accident are taken into account.

The general comparison of the main parameters between the both ATHLET-CD and MELCOR showed that qualitatively good agreement is reached for the pressure in the primary circuit (Fig. 9). Differences between the both codes are more visible in the late phase of the accident scenario, for example, when the melt starts to relocate to the RPV lower plenum. Further work is necessary to analyse more deeply the differences especially in the late phase of the accident progression.

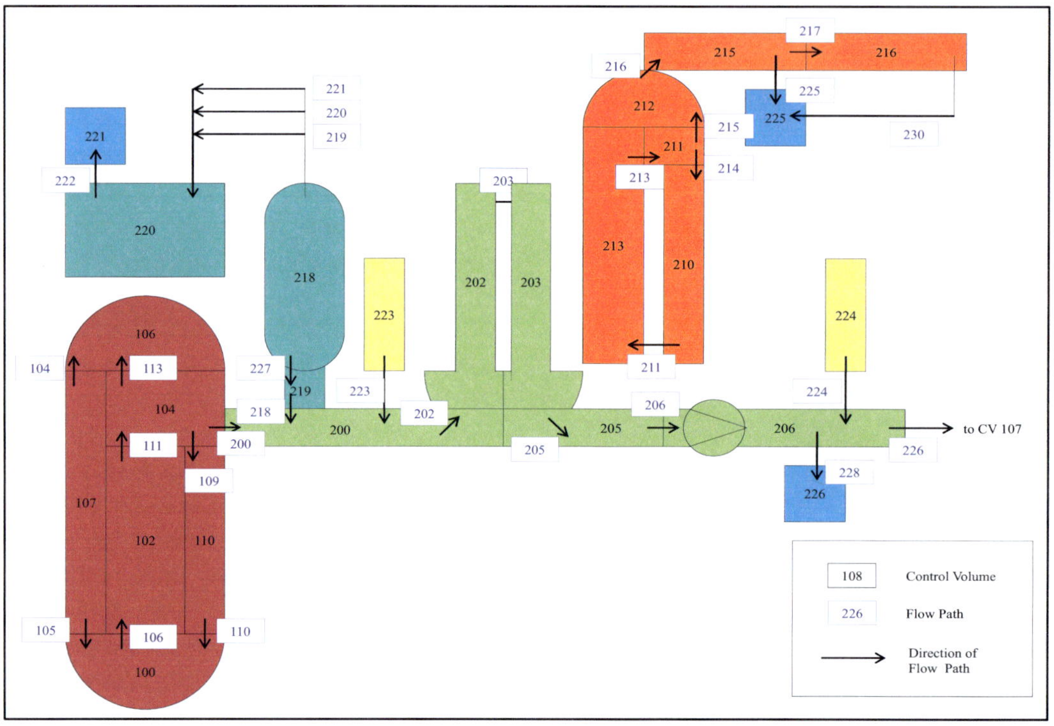

Fig. 8: Nodalization of the generic KONVOI MELCOR model

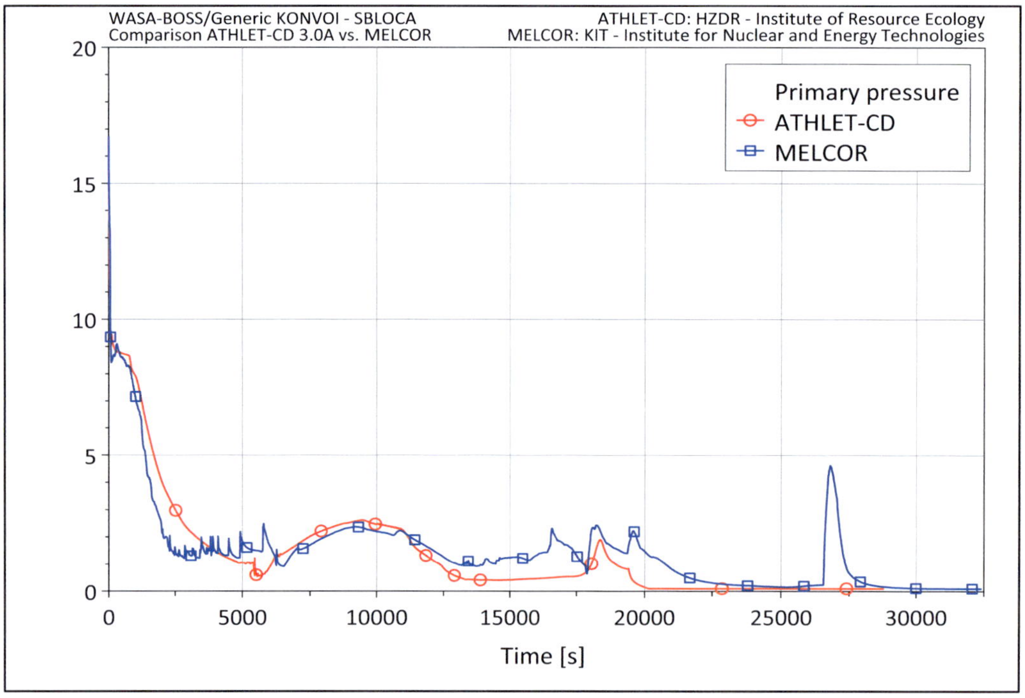

Fig. 9: Primary circuit pressure calculated by MELCOR and ATHLET-CD

References

[1] Dietrich, P. (2016). Expansion of the Severe Accident Code MELCOR by Coupling External Models (KIT Scientific Reports; 7726). Dissertation. KIT Scientific Publishing, Karlsruhe. doi:10.5445/KSP/1000061998

[2] Fischer, M.; Bechta, S. V.; Bezlepkin, V. V.; Hamazaki, R.; Miassoedov, A. (2016). Core Melt Stabilization Concepts for Existing and Future LWRs and Associated Research and Development Needs. Nuclear technology, 196 (3), 524-537. doi:10.13182/NT16-19

[3] Foit, J.J., Influence of cooling on spreading. KAERI, Gyeongju, KR, October 12, 2016.

[4] Foit, J.J. (2016), From Experiment to Reactor. Nuclear Power Engineering Center (NUPEC), Tokyo, J, October 20, 2016.

[5] Foit, J.J. (2016). Mocka versus CCI experiments. Japan Atomic Power Company, Tokyo, J, October 21, 2016.

[6] Foit, J. J. (2016). MOCKA experiments on LCS concrete with and without rebars. International Congress on Advances in Nuclear Power Plants (ICAPP 2016), San Francisco, CA, April 17-20, 2016. Proceedings on CD-ROM, 1827-1831, American Nuclear Society, LaGrange Park (IL).

[7] Gaus-Liu, X. (2016). Summary of LIVE experiments and simulation works. Korea Institute of Nuclear Safety, Daejeon, Korea, October 13, 2016.

[8] Gaus-Liu, X.; Miassoedov, A. (2016). Heat transfer of stratified melt pool in LWR lower head under external cooling and different upper boundary conditions. 11th International Topical Meeting on Nuclear Thermal-Hydraulics, Operation and Safety (NUTHOS-11), Gyeongju, Korea, October 9-13, 2016.Proceedings on USB-Stick, Paper No.N11P0086.

[9] Gaus-Liu, X.; Miassoedov, A.; Gabriel, S. (2016). Review of experimental studies on the heat transfer behaviour of volumetrically-heated pool with different boundary conditions and the influence of crust formation. 24th International Conference on Nuclear Engineering (ICONE24), Charlotte, N.C., June 26-30, 2016. Proceedings on CD-ROM, ICONE24-60268, ASME, New York (NY).

[10] Journeau, C.; Bouyer, V.; Cassiaut-Louis, N.; Fouquart, P.; Piluso, P.; Ducros, G.; Gossé, S.; Guéneau, C.; Quaini, A.; Fluhrer, B.; Miassoedov, A.; Stuckert, J.; Steinbrück, M.; Bechta, S.; Kudinov, P.; Hozer, Z.; Guba, A.; Manara, D.; Bottomley, D.; Fischer, M.; Langrock, G.; Schmidt, H.; Kiselova, M.; Ždarek, J. (2016). SAFEST Roadmap for Corium Experimental Research in Europe. 2016 24th International Conference on Nuclear Engineering, Charlotte, North Carolina, USA, 26 June 2016, V004T14A019, ASME, New York (NY). doi:10.1115/ICONE24-60916

[11] Komlev, A.; Villanueva, W.; Bechta, S.; Miassoedov, A.; Gaus-Liu, X.; Carenini, L.; Fichot, F. (2016). Heat flux distribution in stratified corium pools. International Workshop 'In-Vessel Corium Retention Strategy: Status of Knowledge and Perspectives', Aix-en-Provence, F, June 6-7, 2016.

[12] Miassoedov, A. (2016). SAFEST Project: Towards pan-European Lab on Corium Behavior in Severe Accidents. Main Objectives and R&D Priorities. NUC Workshop: Accident Management: Emerging R&D Needs and Approaches, Raleigh, NC, August 23-25, 2016.

[13] Miassoedov, A.; Albrecht, G.; Meyer, L.; Meignen, R. (2016). Ex-vessel fuel-coolant interaction experiment in the DISCO facility in the LACOMECO project. Journal of Nuclear Science and Technology, 53 (5), 639-646. doi:10.1080/00223131.2015.1127785

[14] Miassoedov, A.; Gaus-Liu, X.; Cron, T. (2016). Study of melt formation and behavior of a two-component debris bed in LIVE facility. Transactions of the American Nuclear Society, 115, 865-868.

[15] Miassoedov, A.; Gaus-Liu, X.; Cron, T. (2016). Study of melt formation and behavior of a two-component debris bed in LIVE facility. American Nuclear Society Winter Meeting, Las Vegas, NV, November 6-10, 2016.

[16] Miassoedov, A.; Gaus-Liu, X.; Su, G.; Zhang, Y.; Tian, W.; Qiu, S. (2016). COPRA and LIVE tests on transient and steady-state heat transfer in a melt pool. International Workshop 'In-Vessel Corium Retention Strategy: Status of Knowledge and Perspectives', Aix-en-Provence, F, June 6-7, 2016.

[17] Miassoedov, A. (2016). Overview of LWR Servere Accident Research Activities at the Karlsruhe Institute of Technology. Nuclear Technology Public Seminar: Recent Severe Accident Research and Development in Europe, Hong Kong, CHN, October 27, 2016.

[18] Miassoedov, A.; Gaus-Liu, X.; Bechta, S.; Komlev, A.; Carenini, L.; Fichot, F. (2016). Heat Transfer in Homogeneous and Stratified Melt Pools in the Lower Head of a Reactor Pressure Vessel. Technical Meeting on Phenomenology and Technologies Relevant to In-Vessel Melt Retention and Ex-Vessel Corium Cooling, Shanghai, CHN, October 17-21, 2016.

[19] Missoedov, A.; Gaus-Liu, X.; Bechta, S.; Komlev, A.; Carenini, L.; Fichot, F. (2016). Assessment of available experimental data used for IVMR demonstration and new complimentary experiments. IAEA Meeting: In-Vessel Melt Retention (IVMR) Strategies for High Power Reactors, Vienna, A, September 27, 2016.

[20] Sanchez, V.; Miassoedov, A.; Steinbrueck, M.; Tromm, W. (2016). KIT multi-physics tools for the analysis of design and beyond design basis accidents of light water reactors. Kerntechnik, 81 (2), 145-148. doi: 10.3139/124.110687

Group: KArlsruhe Liquid metal LAboratory

Concentrated Solar Power (CSP)

Jonathan Flesch, Klarissa Niedermeier, Lucca Marocco, Thomas Wetzel & KALLA-Team

Liquid metals have been proposed in the past as high temperature heat transfer media in concentrating solar power (CSP) systems. IKET contributes experience in liquid metal technology to investigate the performance of liquid metals in CSP within a Helmholtz-Alliance joint research project called LIMTECH. The overall goals of this project are planning, design, construction and operation of a small concentrating solar power system in the 10 kW thermal range (named SOMMER) using liquid metal as heat transfer fluid for re-gaining operation experience, validating the design methodology and providing a complete design concept for a pilot CSP plant based on liquid metal technology, up to evaluation of O&M cost and levelized cost of electricity.

Scientific achievements

The liquid metal loop at KIT is under construction. The measurement system for the incident heat flux on the receiver has been further developed and will be used in the upcoming experimental campaign [3, 4]:

The solar receiver of the SOMMER plant shall be tested at an average heat flux of 1 MW/m^2 and a total power of 10 kW. Ray tracing simulations of the flux distribution on the receiver revealed that local peaks of up to 2.5 MW/m^2 had to be expected directly in the focal point of the concentrating parabolic mirror. An uncooled piece of carbon steel easily begins to melt after a few seconds when exposed to such high fluxes. The heliostat mirror, shown in Fig. 1 that was obtained for the SOMMER project is slightly canted[1] for its operation in a solar tower power facility, resulting in two peaks rather than one with the benefit of a more evenly distributed flux on the receiver.

In order to determine the thermal efficiency of the solar receiver in the SOMMER facility the solar power needs to be determined before it is absorbed, reflected and re-radiated by the receiver surface. Commonly used methods for the measurement of the flux such as a moving Lambertian target were found not to be suitable[2]. An existsing concept using a set of moving heat flux microsensors (HFM) was adopted and adjusted (Fig. 2) for the service in the SOMMER facility with the improvement that only a single sensor is required instead of several while at the same time increasing the resolution of the obtained image. 10000 values can be measured on a plane of 100 mm by 100 mm. The heat flux microsensor is already calibrated by the manufacturer at purchase. The sensor used in SOMMER will be coated with Pyromark 2500, a coating with accessible and well-published spectral properties. The HFM is mounted on the tip of a double-pivoted rotating arm, which moves the sensor on a circular path. A linear motion component is superimposed by a linear actuator.

---

1 Canting: The mirror facets are not perfectly arranged in a plane but tilted slightly so that the projected images of all facets overlap in the receiver plane. This allows for more precisely hitting the receiver with the reflected sunlight. The design distance of the heliostat mirror is, however, larger than the distance to the receiver obtained in the SOMMER facility.

2 Why?: These systems require a reference flux gauge mounted in the measuring plane. In SOMMER the flux is measured a distance in front of the receiver and the rays approach the receiver from within a view angle of 90° horizontally and vertically. As a result the reference acquires values not valid in the plane of measurement.

Fig. 1: Heliostat mirror of the SOMMER project, reflected in the lower left facet: the concentrating parabolic mirror.

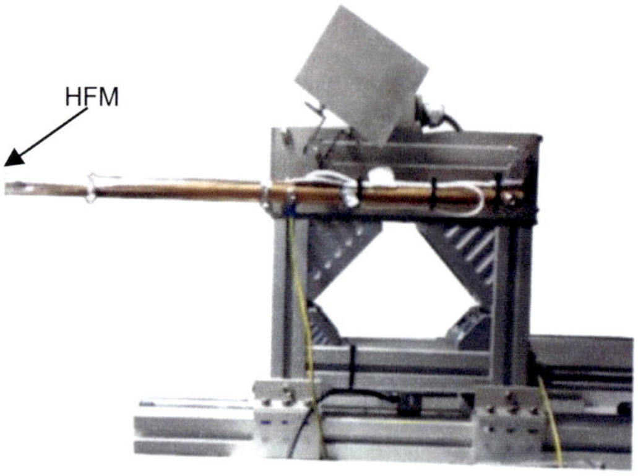

Fig. 2: Device for the measurement of solar flux and its distribution on the receiver surface.

As a result the sensor is moved across the receiver plane and constantly scanning the local flux values. Proper interpolation between those discrete samples generates a surface which upon integration allows to calculate the incident power.

In 2016 the device was tested in the solar furnace of the SOMMER facility for the first time. The measured flux distribution (on the right of Fig. 3) is quite similar to the previously simulated distribution on the left.

During a later test at DNI = 770 W/m² the predicted peak flux was nearly reached (Fig 4) while the shutter blinds let only pass 80% of the light onto the parabolic mirror.

The expected uncertainty of the value of total power measured is below 6%. The images indicate that the design power of the SOMMER plant as well as the expected peak and average flux can be obtained during the upcoming experimental campaign. Once some minor mechanical improvements of the system are completed the device can be used in the SOMMER project.

In addition to the progress in the measurement of solar flux a detailed study on the applicability of liquid metals as thermal storage fluid and the possible storage configurations has been done [1]. It emerges that the direct one tank system with filler material is a very promising option, together with the possibility of including phase-change-material (PCM) capsules. Further research on this topic is currently ongoing.

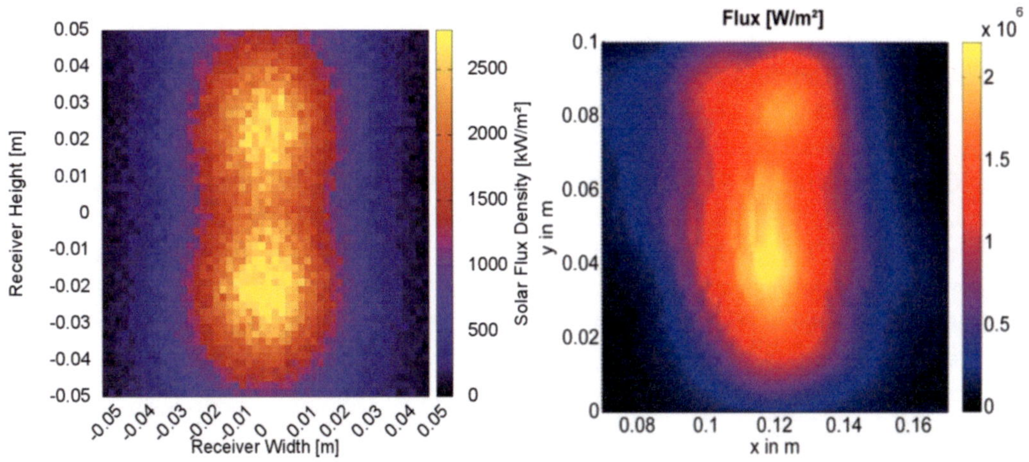

Fig. 3: Left: ray-tracing simulation of the optical system at DNI = 750 W/m² and solar noon; Right: measured distribution at one hour after solar noon at DNI = 770 W/m² and partly closed shutter (at 7 kW total power).

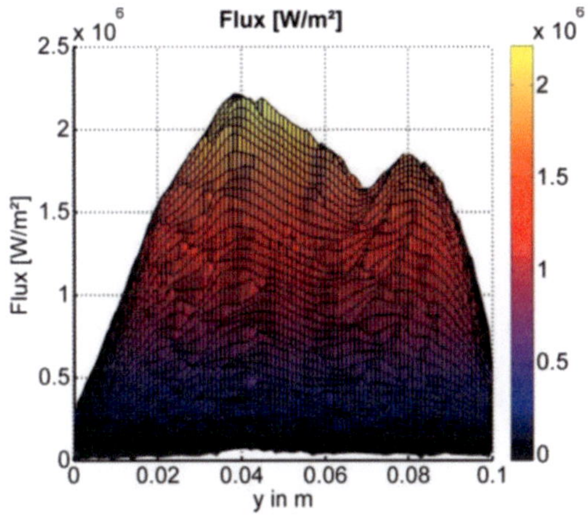

Fig. 4: Vertical cut of the flux distribution on the receiver at DNI = 770 W/m² and 80% opening of the shutter blinds

A comparison between 1D-models and 3D-models for the design of the receiver panel has been done. The results highlight the necessity of knowing the distribution of the Nusselt number on the non-homogeneously heated surface, in order to properly predict the wall temperature with the 1D simplification.

A detailed numerical analysis of turbulent aided mixed convection to a liquid metal flowing in an annulus has been done [2]. It has been shown that the results and conclusions obtained for medium-to–high Prandtl number fluids cannot be extrapolated to the case of liquid metal flows.

Networking

A collaboration with the Australian National University (ANU) has been established. They are studying a CSP system with sodium as heat transfer fluid.

References

[1] Niedermeier, K.; Flesch, J.; Marocco, L.; Wetzel, T. (2016). Assessment of thermal energy storage options in a sodium-based CSP plant. Applied thermal engineering, 107, 386–397. doi:10.1016/j.applthermaleng.2016.06.152

[2] Marocco, L.; Alberti di Valmontana, A.; Wetzel, Th.; Numerical investigation of turbulent aided mixed convection of liquid metal flowing through a concentric annulus, Int. Jou. Heat Mass Transf. 105 (2017) 479-494.

[3] Albrecht, C.; Messung der Leistung konzentrierter Solarstrahlung in einem Sonnenofen, M.Sc. thesis. Karlsruhe Institute of Technology (KIT), 2016

[4] De Geus, J.; System zur Flussdichtemessung in einem Sonnenofen, M.Sc. thesis. Karlsruhe Institute of Technology (KIT), 2015

Read more:

[5] Abánades, A.; Rathnam, R. K.; Geißler, T.; Heinzel, A.; Mehravaran, K.; Müller, G.; Plevan, M.; Rubbia, C.; Salmieri, D.; Stoppel, L.; Stückrad, S.; Weisenburger, A.; Wenninger, H.; Wetzel, T. (2016). Development of methane decarbonisation based on liquid metal technology for CO2-free production of hydrogen. International Journal of Hydrogen Energy / Special Issue on Progress in Hydrogen Production and Applications, 41 (19), 8159–8167. doi:10.1016/j.ijhydene.2015.11.164

[6] Geißler, T.; Abánades, A.; Heinzel, A.; Mehravaran, K.; Müller, G.; Rathnam, R. K.; Rubbia, C.; Salmieri, D.; Stoppel, L.; Stückrad, S.; Weisenburger, A.; Wenninger, H.; Wetzel, T. (2016). Hydrogen production via methane pyrolysis in a liquid metal bubble column reactor with a packed bed. The chemical engineering journal, 299, 192-200. doi: 10.1016/j.cej.2016.04.066

[7] Marocco, L.; Cammi, G.; Flesch, J.; Wetzel, T. (2016). Numerical analysis of a solar tower receiver tube operated with liquid metals. International Journal of Thermal Sciences, 105, 22–35. doi:10.1016/j.ijthermal sci.2016.02.002

[8] Pacio, J.; Daubner, M.; Fellmoser, F.; Litfin, K.; Wetzel, T. (2016). Experimental study of heavy-liquid metal (LBE) flow and heat transfer along a hexagonal 19-rod bundle with wire spacers. Nuclear Engineering and Design, 301, 111-127. doi:10.1016/j.nucengdes. 2016.03.003

[9] Pacio, J.; Litfin, K.; Wetzel, T.; Knebel, J. (2016). Thermal-hydraulic experiments at KALLA in support of ADS. 3rd International Workshop on Technology and Components of Accelerator-Driven Systems, Mito, J, September 6-9, 2016.

[10] Schumm, T.; Frohnapfel, B.; Marocco, L. (2016). Numerical simulation of the turbulent convective buoyant flow of sodium over a backward- facing step. Journal of physics / Conference Series, 745 (3), Art. Nr.: 032051. doi:10.1088/1742-6596/745/3/032051

[11] Wetzel, T.; Daubner, M.; Fellmoser, F.; Litfin, K.; Marocco, L.; Pacio, J. (2016). Experimental activities at KALLA on heavy-liquid metal heat transfer for fast reactors. International Congress on Advances in Nuclear Power Plants (ICAPP 2016), San Francisco, CA, April 17-20, 2016. Proceedings on CD-ROM, 1294-1303, American Nuclear Society, LaGrange Park (IL).

Group: Accident Management Systems

Installation of the JRodos System in Ukraine and China

Dmytro Trybushnyi, Claudia Landman, Yu Li, and Wolfgang Raskob

Introduction

For almost three decades, the European Commission has promoted and supported the development of a non-commercial computer-based decision support system for off-site emergency management and rehabilitation issues after nuclear accidents, RODOS, and its Java based successor since 2010, JRodos. The final version of RODOS (PV 7.0) and current versions of JRodos are installed in national emergency centres of several European countries, such as Austria, Germany, Bulgaria, Czech Republic, Finland, Hungary, The Netherlands, Poland, Portugal, Russia, Spain, Slovakia, Slovenia, Switzerland and outside Europe in Hong Kong. Installation in Central Europe and the former Soviet Union has been achieved with support from the European Commission's ECHO, PHARE and TACIS programs, respectively.

In 2016, two JRodos installations in China and Ukraine, supported by the European Commission, have been completed.

Installation approach

The Installation is typically subdivided into several tasks. First, hard and software components for the emergency centre operating JRodos have to be defined. In a second step, the connection to data providers has to be established. Data provider are typically national weather services with their weather forecast data, the power plants with their radiological and meteorological data and finally national monitoring networks that can be linked to the system. Finally the collected data has to be integrated into the system which requires on the one hand side adaptation of the data format and on the other hand provision of data links to obtain the real-time monitoring data from the power plant (typically every 10 or 30 minutes) and the numerical weather forecast data on a regular basis (at least once per day). The key task is the data collection itself. The data to be collected can be subdivided into

- plant and site characteristics for each NPP in the country,
- map data such as buildings, vegetation, land use, roads and borderlines
- statistical data such as demographic, agricultural production and economic data,
- relevant parameters of food-chain and ingestion models,
- relevant data for the hydrological model chain,
- intervention criteria for countermeasures and emergency plans, and
- translation of the JRodos user interface into the national language.

In particular the collection of parameters for the foodchain and dose module takes most of the resources foreseen in such contracts. As JRodos is delivered with a default dataset covering the whole world, test installations can be performed at any time to support the data collection.

Once the data collection is complete, a final installation is carried out which is further tested in its operational environment by applying test procedures that were specifically developed for that installation. The whole process is accompanied by training course and methodological support for the organisation operating the emergency centre.

Objectives

Ukraine

The aim of this project was to improve the technical capabilities of the Ukrainian government for the detection, planning and practical management and initiation of countermeasures for protection of personnel, population and environment in the case of nuclear accident at any Ukrainian NPP. This should be achieved by installing the JRodos decision support system operational within an operational emergency centre in Kiev (Centre of Prognosis of Radiation Accident Consequences “CPRAC”) and in attrition at the regulator, at the main emergency centre of SE NNEGC Energoatom and all four NPPs.

China

The objective here was to customise the JRodos decision support system for emergency management for use in China, install it in the National Nuclear Emergency Response Technical Assistance Centre (NNERTAC), provide training on its use and verify its operability. The functional test of the system should be demonstrated at different Chinese NPPs.

Work performed and progress achieved

General

Even if the objectives for both installations in Ukraine and China looked similar, the methodology applied was very different. In Ukraine, a “typical” installation process was performed. KIT was responsible for data collection, customisation of data and implementation of the data into the JRodos system. A local subcontractor and the beneficiaries supported the work. In China, KIT had only a supervising role and no access to the JRodos installation. Data collection, customisation and implementation was solely performed by the Chinese beneficiaries. This caused a high burden to KIT in documentation and explanation of tasks typically performed by KIT experts. In this respect, the installation in China was very demanding and required many personal contacts of the team with counterparts in China. Only during visits, the installation progress could be determined and problems encountered solved.

Tasks

The following highlights result from both installations. The most demanding task is the customisation of the food chain and dose module FDMT. To operate FDMT in all areas of the world, so called “radioecological regions” were introduced. A radioecological region is defined as a region with similar vegetation and dietary characteristics. This requires interaction with agricultural services in that particular country. Typically a country is subdivided into 5 regions as was done for Ukraine (Fig. 1).

Fig. 1: Subdivision of Ukraine in 5 radioecological regions

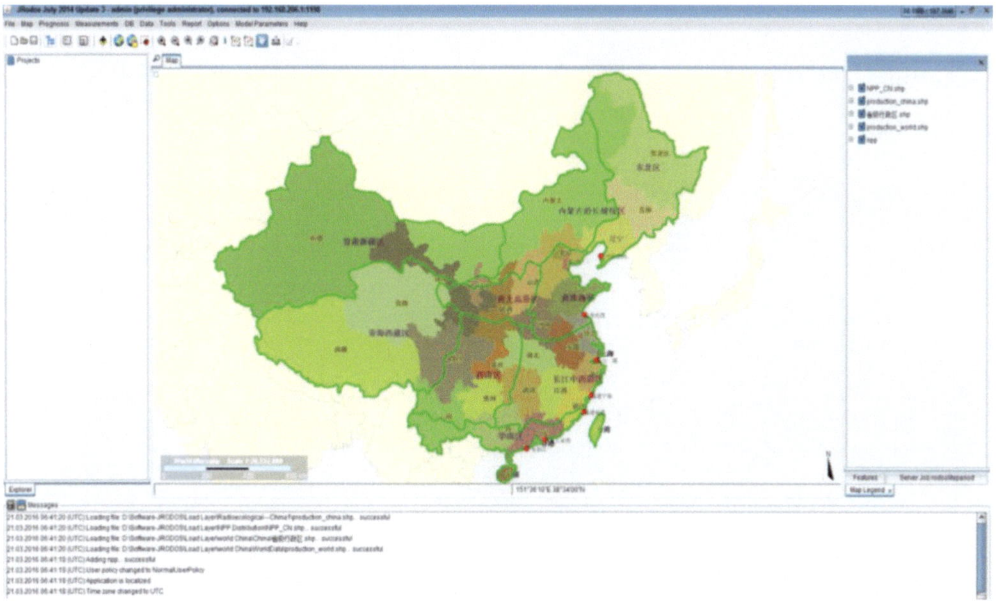

Fig. 2: Subdivision of China in 9 radioecological regions

For China, the climatic regions and vegetation characteristics necessitated a larger number of regions. Therefore, the FDMT model had to be expanded and can now accommodate up to 25 different regions. China was subdivided into 9 regions (Fig. 2). In addition, the KIT team had to add rice as new foodstuff into FDMT as rice is one of the dominant foods in China.

Besides the customisation of the FDMT model, the integration of real-time data from the nuclear power plants (NPPs) as well as the numerical weather data from the national weather provider are the keys of success for such an installation. With the weather data from the NPPs, the atmospheric transport and dispersion models can be operated in their prognostic and diagnostic modes. Fig. 3 and 4 show examples from Ukraine and China, respectively. Numerical weather data is delivered operationally to the emergency centres. As fallback, weather data from the US are freely available. The US National Oceanic and Atmospheric Administration NOAA operates four different NOMADS Servers that host National Centres for Environmental Prediction (NCEP) model products. The NCEP service can be accessed via the following web page: www.ncep.noaa.gov/.

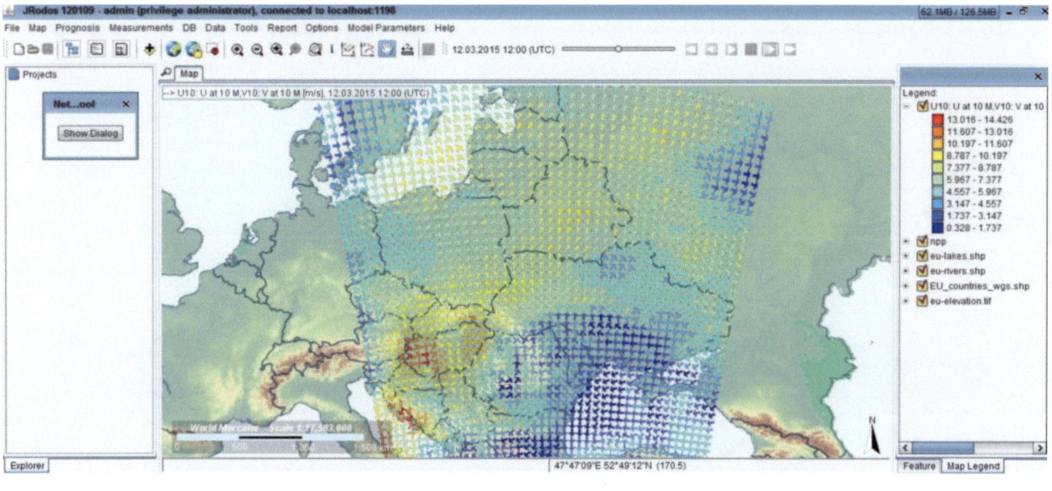

Fig. 3: Total area covered by wind field model for Ukraine

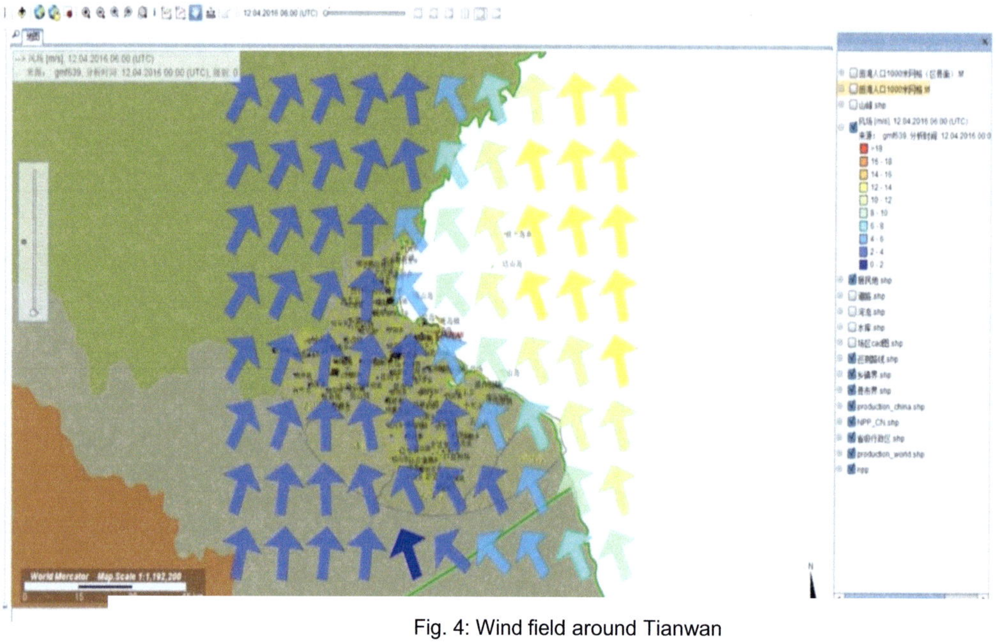

Fig. 4: Wind field around Tianwan

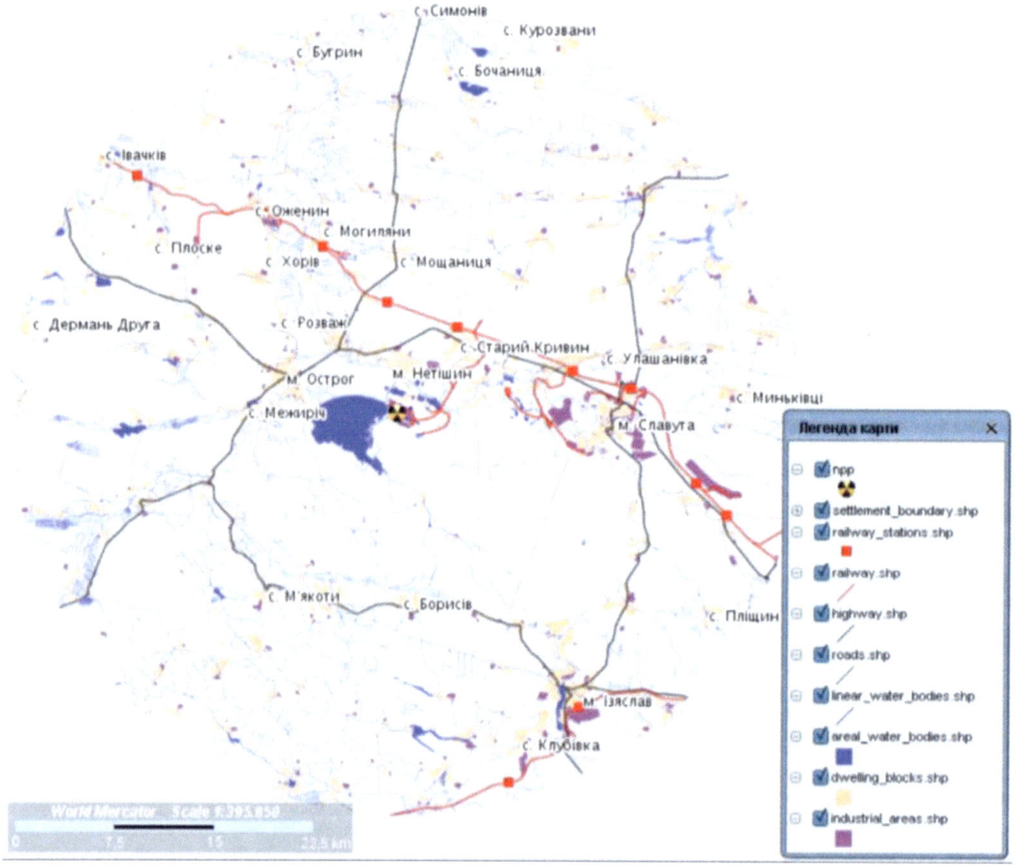

Fig. 5: Map information in the 30 km cycle around an NPP

It offers a forecast of up to several days with updates every 6 hours. The data cover the whole world with a resolution of either 0.5 or 0.25 degrees.

For visualisation, specific maps around the NPPs have to be prepared. In Ukraine, a company extracted the relevant data from information available and KIT integrated them into the JRodos system (see Fig. 5). For China, the beneficiaries there collected the necessary information and integrated them into the formats required by JRodos (see Fig 6.)

The final task is the translation of the user interface and partially the map results into the national language (see Fig. 7 and Fig. 8). With the JRodos system, it is possible to use any character set as is seen for Ukraine and China.

Lessons learned

There are two aspects which are most important in the installation process of the JRodos system in a national context. First, the contractor has to be flexible in case the beneficiary has a request which was not foreseen either in the contract or not foreseen to be realised at a particular time. Second, in particular the data collection process for the foodchain and dose module FDMT is not trivial and sometimes the first attempt to find good data sources is not successful. In this respect, the time frame defined at the inception meeting should be regarded as guidance document, however, if justified, deviations should be possible.

The local subcontractor is extremely important as he knows the local conditions and the organisational structures of the beneficiaries and national organisations which have to be approached for data delivery. A careful selection of such a subcontractor is mandatory and the 30% of resources allocated for that task is a good investment.

The Chinese installation was special, as KIT was only supervising the data collection and implementation. This made the direct interaction between beneficiaries and KIT mandatory and regular meetings in China were foreseen. However, supervising of a task without access to an installation – except during the visits – was nearly impossible and only successful as all parties were highly interested in a positive outcome of the installation process. Finally, the selection of an interpreter who knows the subject is also a key for success as otherwise KIT would have had severe problems in communicating technical details during the meetings in China.

Fig. 6: Map information around Tianwan

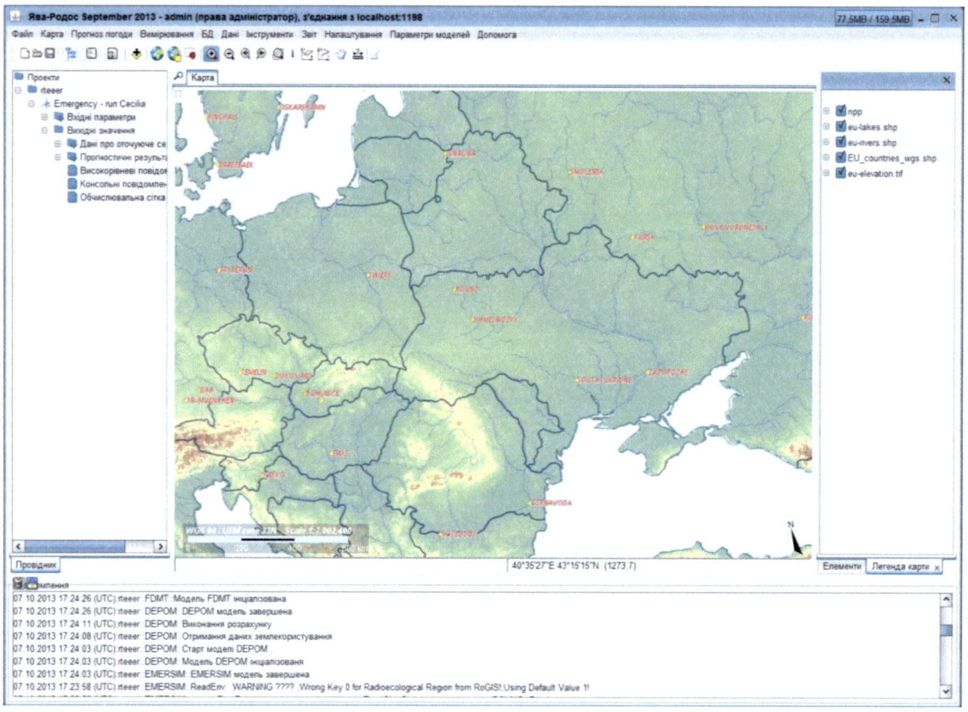

Fig. 7: Ukrainian user interface

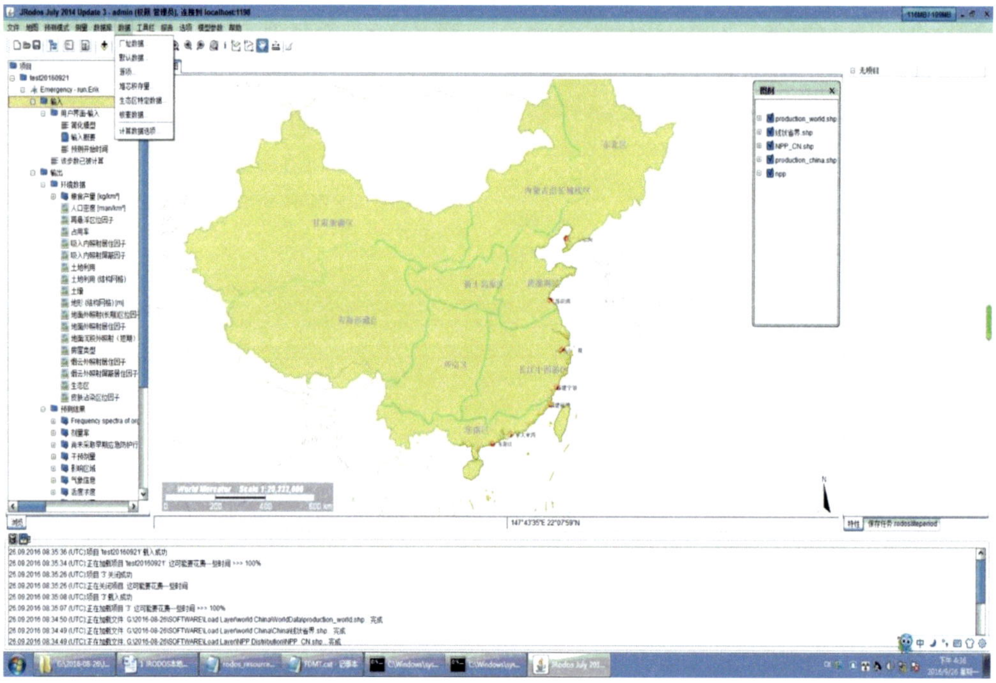

Fig. 8: Chinese user interface

Reviewing both approaches, the full integration of KIT in the data collection and integration as in Ukraine and the supervising approach which was carried out in China, one has to state that the first is more appropriate for a successful implementation of JRodos to national conditions.

Conclusions

JRodos is a decision support system for offsite emergency management following releases of radioactive material into the environment. It contains detailed simulation models for predicting and analysing the resulting contamination, health and economic consequences. JRodos is a non-commercial system with an active user community that influences system extensions and development trends. The default database is for use in Central Europe. A world-wide database and the supported coupling to a set of globally applicable meteorological weather forecast data allows general application for any point worldwide. Furthermore, inherent features and tools allow the adaptation of models, data bases, and the user interface to national conditions and user preferences.

The adaptation to national conditions in countries operating the system in national emergency centres requires collecting data for several of the simulation models as the default database might be not applicable. KIT supports this process either by actively collecting the data itself and implementing them or by supervising the whole process.

As a consequence each installation has improved the system with respect to functionalities and stability – acceptance and verification tests had to be developed for each national installation. Furthermore, the documentation of the system was much improved aiming to minimise the effort of KIT in explanations and training. In this respect also self-learning training courses were developed which support the first installation of the system without having participated in a detailed training course. The integration of these countries into the RODOS User Group (RUG) is very important. The aim of the RUG is to exchange information about the use of JRodos and to discuss the way of further development of the system. With about 15 organisations supporting the maintenance and further development of the system, the RUG is an important factor for the stability and sustainability of JRodos.

References

[1] Andronopoulos, S.; Ievdin, I.; Kovalets, I.; Anulich, S.; Trybushnyi, D. (2016). New functionalities developed in the NERIS-TP project regarding meteorological data used by Decision Support Systems. Radioprotection, 51 (HS1, S, SI), S13-S16. doi:10.1051/radiopro/2016004

[2] Andronopoulos, S.; Schichtel, T.; Efthimiou, G.; Bartzis, J. G. (2016). Updates of the atmospheric dispersion models inside the Local Scale Model Chain of RODOS regarding particles. Radioprotection, 51 (HS2), 101–103. doi:10.1051/radiopro/2016041

[3] Charnock, T.; Landman, C.; Trybushnyi, D.; Ievdin, I. (2016). European model for inhabited areas - ERMIN 2. Radioprotection, 51 (HS1, S, SI), S23-S25. doi:10.1051/radiopro/2016006

[4] Duran, J.; Duranova, T.; Bohunova, J.; Kovalets, I.; Trybushnyi, D. (2016). Fast and simple method with NPP fence measurements for source term estimation and its integration in JRODOS. Radioprotection, 51 (HS2), 117–119. doi:10.1051/radiopro/2016045

[5] Duran, J.; Duranova, T.; Bohunova, J.; Kovalets, I.; Trybushnyi, D. (2016). Fast and simple method with NPP fence measurements for source term estimation and its integration in

JRODOS. PREPARE Dissemination Workshop "Innovative integrative tools and platforms", Bratislava, SK, January 20-22, 2016.

[6] Landman, C.; Raskob, W.; Trybushnyi, D.; Ievdin, I. (2016). An application example for the new ICRP screening tool of JRodos. Radioprotection, 51 (HS1, S, SI), S17-S22. doi:10.1051/radiopro/2016005

[7] Landman, C.; Raskob, W.; Trybushnyi, D.; Ievdin, I. (2016). Creating and running worldwide accident scenarios with JRodos. Radioprotection, 51 (HS1, S, SI), S27-S30. doi:10.1051/radiopro/2016007

[8] Montero, M.; Trueba, C.; Duranova, T.; Bohunova, J.; Raskob, W. (2016). Operational procedures of analytical platform. Radioprotection, 51 (HS2), 191–193. doi:10.1051/radiopro/2016071

[9] Montero, M.; Trueba, C.; Duranova, T.; Bohunova, J.; Raskob, W. (2016). Operational procedures of analytical platform. PREPARE Dissemination Workshop "Innovative integrative tools and platforms", Bratislava, SK, January 20-22, 2016.

[10] Müller, T. O. (2016). Technical realization of the analytical platform. Radioprotection, 51 (HS2), Springer International Series in Operations Research 235, 181–183. doi:10.1051/radiopro/2016068

[11] Müller, T. O. (2016). Technical realization of the analytical platform. PREPARE Dissemination Workshop "Innovative integrative tools and platforms", Bratislava, SK, January 20-22, 2016.

[12] Münzberg, T.; Wiens, M.; Schultmann, F. (2016). Understanding resilience: A spatiotemporal vulnerability assessment of a population affected by sudden lack of food. Advances in Managing Humanitarian Operations. Ed.: C.W. Zobel, 257-280, Springer International Series in Operations Research 235, Cham. doi:10.1007/978-3-319-24418-1

[13] Raskob, W. (2016). The PREPARE project - Innovative integrated tools and platforms for radiological emergency preparedness and post-accident response in Europe. Radioprotection, 51 (HS1, S, SI), S55-S57. doi:10.1051/radiopro/2016014

[14] Raskob, W.; Landman, C.; Trybushnyi, D. (2016). Functions of decision support systems (JRodos as an example): overview and new features and products. Radioprotection, 51 (HS1), S9-S11. doi:10.1051/radiopro/2016015

[15] Raskob, W.; Möhrle, S.; Bai, S. (2016). Knowledge database and case-based reasoning. Radioprotection, 51 (HS2), 185–186. doi:10.1051/radiopro/2016069

[16] Raskob, W.; Möhrle, S.; Bai, S. (2016). Knowledge database and case-based reasoning. PREPARE Dissemination Workshop "Innovative integrative tools and platforms", Bratislava, SK, January 20-22, 2016.

[17] Raskob, W.; Möhrle, S.; Bai, S.; Müller, T. (2016). The analytical platform of the PREPARE project. 14th Congress of the International Radiation Protection Association (IRPA), Cape Town, ZA, May 9-13, 2016.

[18] Raskob, W.; Möhrle, S.; Bai, S.; Müller, T. (2016). Overview and applicability ot the analytical platform. PREPARE Dissemination Workshop "Innovative integrative tools and platforms", Bratislava, SK, January 20-22, 2016.

[19] Raskob, W.; Möhrle, S.; Bai, S.; Müller, T. (2016). Overview and applicability of the analytical platform. (T. Duranova, Raskob, W. & Schneider, T., Hrsg.) Radioprotection, 51 (HS2), 179–180. doi:10.1051/radiopro/2016067

[20] Raskob, W.; Schneider, T.; Gering, F.; Charron, S.; Zhelezniak, M.; Andronopoulos, S.; Heriard-Dubreuil, G.; Camps, J. (2016). Nuclear and radiological preparedness: The

achievements of the European research Project PREPARE. 14th Congress of the International Radiation Protection Association (IRPA), Cape Town, ZA, May 9-13, 2016.

[21] Raskob, W.; Schneider, T.; Gering, F.; Charron, S.; Zheleznyak, M.; Andronopoulos, S.; Heriard-Dubreuil, G.; Camps, J. (2016). Innovative integrative tools and platforms. Key results of the PREPARE European Project. Radioprotection, 51 (HS2), 59–61. doi:10.1051/radiopro/2016032

[22] Raskob, W.; Schneider, T.; Gering, F.; Charron, S.; Zheleznyak, M.; Andronopoulos, S.; Heriard-Dubreuil, G.; Camps, J. (2016). Innovative integrative tools and platforms: Key results of the PREPARE European Project. PREPARE Dissemination Workshop "Innovative integrative tools and platforms", Bratislava, SK, January 20-22, 2016.

[23] Schneider, T.; Lafage, S.; Bardelay, J.; Duranova, T.; Gallego, E.; Gering, F.; Hardeman, F.; Heriard-Dubreuil, G.; Murith, C.; Oughton, D.; Raskob, W. (2016). NERIS: European platform on preparedness for nuclear and radiological emergency response and recovery - Activities and perspectives. Radioprotection, 51 (HS1), S5-S8. doi:10.1051/radiopro/2016003

[24] Turcanu, C.; Crouäil, P.; Duranova, T.; Camps, J.; Schneider, T.; Raskob, W. (2016). Training courses on emergency preparedness, response and recovery: theory, practice and application of newly developed tools. Radioprotection, 51, 171–173. doi:10.1051/radiopro/2016065

[25] Turcanu, C.; Crouäil, P.; Duranova, T.; Camps, J.; Schneider, T.; Raskob, W. (2016). Training courses on emergency preparedness, response and recovery: theory, practice and application of newly developed tools. PREPARE Dissemination Workshop "Innovative integrative tools and platforms", Bratislava, SK, January 20-22, 2016.

Group: Hydrogen

Hydrogen Risk Assessment for Nuclear Applications and Safety of Hydrogen as an Energy Carrier

Andrei Denkevits, Benedikt Hoess, Olaf Jedicke, Thomas Jordan, Birgit Kaup, Ke Ren, Alexei Kotchourko, Mike Kuznetsov, Alexander Lelyakin, Frank Prestel, Michael Ranft, Reinhard Redlinger, Maria Stassen, Jianjun Xiao, Zhanjie Xu, Jorge Yanez, Yu Zhang

Introduction

In 2016 the Hydrogen Group continued to develop the in-house CFD HyCodes, GASFLOW and COM3D, and conducted several experimental programs to deepen the understanding of the transient flame acceleration and deflagration-detonation-transition or to provide model validation data. Still the code development is accommodated in the nuclear program NUSAFE and therefore suffers from continuous budget constraints. However, the quite successful commercial licensing of GASFLOW and COM3D provides additional financial resources for the general safety research done at the Hydrogen Group.

The next chapters describe outstanding activities of the group: one highlights CFD code related development, the dynamic grid refinement for COM3D, the other is about experiments investigating the dynamic hydrogen flame behavior in micro-channels.

Dynamic computational grid in COM3D

On 1st December 2016 the new version COM3D 5.0 was released. The new version COM3D is the optimized version and contains lots of new features.

<u>Higher calculation efficiency</u>

First of all, the new version COM3D keeps the same capability in uniform grid simulation as the old version 4.10. But the calculations made by the new version are more efficient than the old version.

COM3D version 5.0 is built upon the MPI parallel C++ library pGrid. The library is developed by the hydrogen safety group. One of the most important techniques contained in the library is its new style of data index system: c-index. The new index system allows the code continuously and efficiently reading and writing of the memory. Additionally, in the coding, new version COM3D uses a more efficient thermo-dynamic library and avoids some unnecessary calculations. Therefore the new version COM3D has better calculation efficiency than the old version, especially for some industrial problems which contain huge amount of solid structures.

Fig. 1 shows the time costs of the COM3D V5.0 and COM3D V4.10 in simulation of 2D turbulent flow. Shown by the comparison, the calculation speed of the new version COM3D is about 2 times faster than the old version.

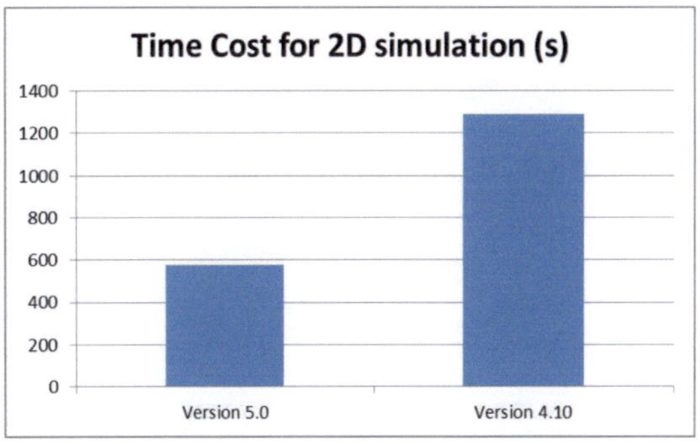

Fig. 1: Time Cost for 2D turbulent flow simulations

Implementation of Local Mesh Refinement:

One of the most important reasons for developing the new library pGrid is to implement the technique local mesh refinement (LMR) to COM3D. Currently the technique is implemented partially in the version 5.0.

The block-structured patch-based local mesh refinement implemented in COM3D is first introduced by Berger and Oliger in 1984. With the help of such technique high resolutions can be achieved locally in the computational domain. Fig. 2 shows a 2D domain in which the LMR is used, compared with uniform grid structure such technique can save large amounts of computational efforts in numerical simulations.

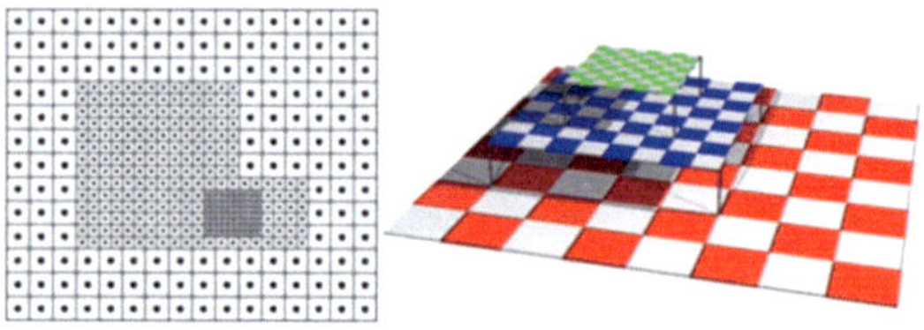

Fig. 2: 2D computational domain with local mesh refinement.

Currently, the technique of local mesh refinement has been implemented in most part of the COM3D: in the solution of Euler equations the LMR is fully implemented; in the solution of the turbulence part (equation right hand side, ERHS), since the complexity of LES models, LMR has been implemented in RANS models only (standard NS model, standard k-e model, RNG k-e model and SST k-o model); in the solution of chemical reaction part LMR has been fully implemented.

To test the implementation of LMR, several verification and validation cases are used.

The first case presented is a 2D convection diffusion flow case. The problem domain is depicted in Fig. 3.

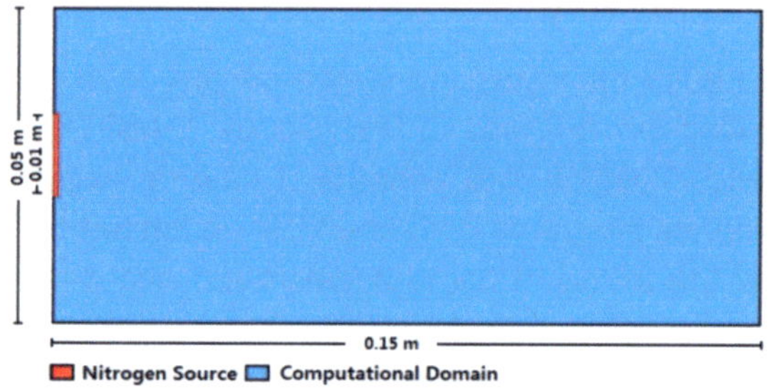

Fig. 3: Domain of 2D convection diffusion verification case

Initially, the computational domain is filled with Oxygen gas and has 5m/s velocity in horizontal direction. Left bound of the computational domain is set as the velocity inflow bound, the middle 0.01m width bound is set as the Nitrogen gas inflow bound and the other parts are set as the Oxygen gas inflow boundary (inflow velocity of the inflow bound is also 5m/s). Right bound of the computational domain is set as the open boundary. The 2D problem has an analytical solution when a constant diffusion coefficient is given.

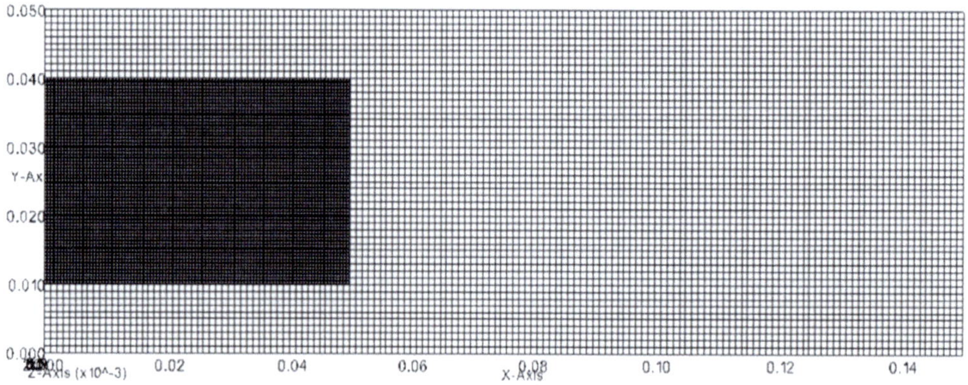

Fig. 4: Grid setting in 2D convection diffusion problem

To control the numerical diffusion, local mesh refinement is used around the Nitrogen Source. Fig. 4 shows the grid setting of the computational domain. In the domain two different resolutions are used: resolution of the base level is 1mm; resolution of the finer refinement level is 0.25 mm. Therefore the most computational efforts are focused on the calculation of Nitrogen diffusion.

Fig. 5 and 6 show the comparison between the numerical solution and analytical solution at time 0.0084s. Fig. 5 shows the comparison of the Nitrogen volume fractions in the central line along the horizontal direction. Fig. 6 shows comparison of the Nitrogen volume fraction in the vertical line 20 mm away from the velocity inflow boundary.

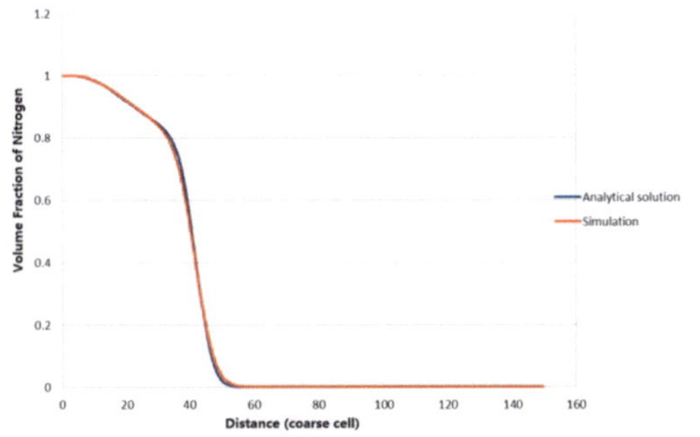

Fig. 5: Comparison of Nitrogen fraction in the central line

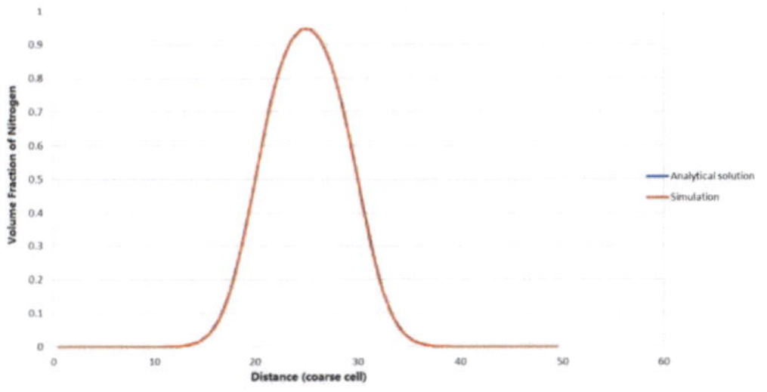

Fig. 6: Comparison of Nitrogen fraction in the vertical line

Both two comparisons show that the calculation with LMR has a quite good compatibility with the analytical solution. In the view of verification the implementation of LMR in code COM3D is successful.

Code to code validation for 2D step flow:

For code development in grid structure, keeping the identical performances as the old version is also quite important. So the second test for the implementation of LMR is code to code validation: the comparison between COM3D version 5.0 and COM3D version 4.10 in simulation of 2D step flow.

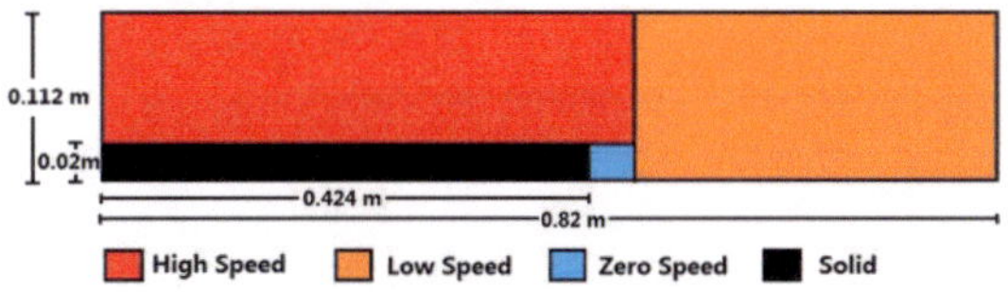

Fig. 7: Domain of 2D step flow

The domain of the 2D step flow problem is divided into 4 parts: the solid region, where the step is located; the high speed region, where the flow has the speed of 44.2m/s in horizontal direction; the low speed region, where the flow has the speed of 38.67m/s in horizontal direction, and the zero speed region, where the gas has no flow speed. Left border of the computational domain is set as the velocity inflow boundary, having the constant 44.2m/s gas inflow, and the right border of the domain is set as the open boundary.

In the 2D problem, due to big a velocity gradient, a big flow vortex is produced after the step. To resolve the big vortex, local mesh refinement is used around the step region. Fig. 8 shows the grid setting of the computational domain: a resolution of 4 mm is used to construct the base level of the computational domain; a refinement of 1 mm is used on the finer refinement level to resolve the region around the step.

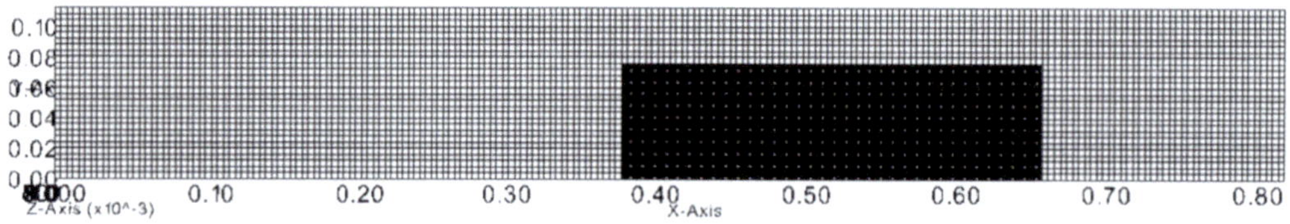

Fig. 8: Grid setting in 2D step flow

In contrast, to achieve the resolution of 1 mm in the step region, a resolution of 1 mm must be used uniformly in the numerical simulation of COM3D version 4.10.

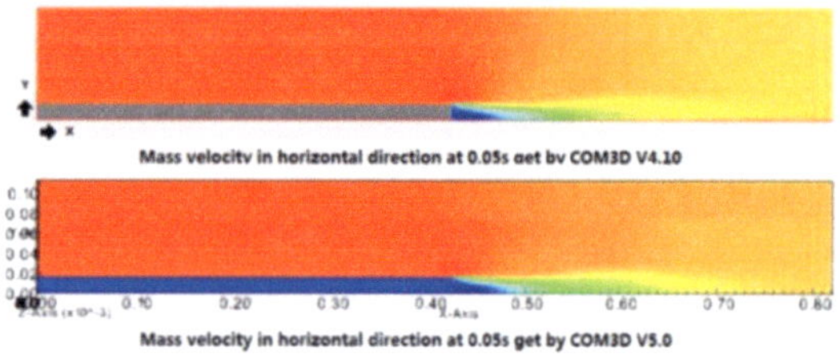

Fig. 9: Mass velocity in horizontal direction at 0.05s

Fig. 9 shows the comparison of mass velocity in horizontal direction at time 0.05s. In the comparison it is not difficult to find that the result achieved by COM3D version 5.0 is the same as the result achieved by COM3D version 4.10. Since the existence of the strong vortex around the step, standard k-e model is used in the right hand part. Fig. 10 shows the comparison of turbulence kinetic energy at time 0.05s. The comparison shows that the COM3D version 5.0 can perform the same as the old version 4.10 in the resolved part (the region where is covered by the resolution 1mm). In general, calculation with LMR in COM3D version 5.0 has identical performances to the old version COM3D.

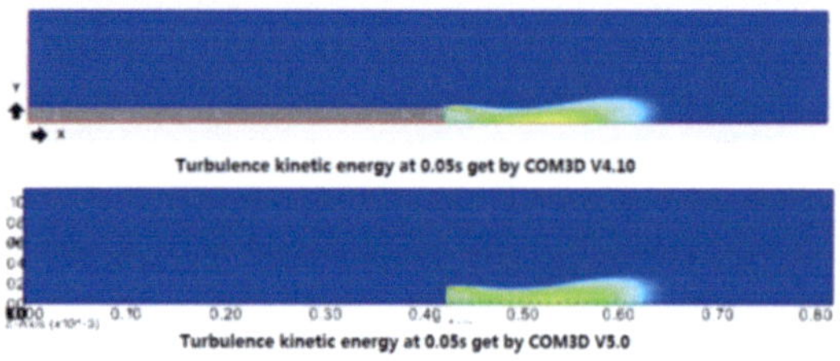

Fig. 10: Turbulence kinetic energy at 0.05s

Experimental investigation of flame acceleration and deflagration-to-detonation-transition in micro channels

Besides model and numerical code development the Hydrogen Group of IKET was continuing its experimental work on fundamental dynamic combustion behavior of premixed systems. Experiments for flame acceleration and deflagration-to-detonation-transition in micro channels have been inspired by industry partners, who integrate fuel cells and electrolysers in their systems. They had observed surprisingly fast combustion regimes in these relatively small structures. Complementing the related flat layer experiments executed in 2015, experiments in a small but long duct, channel respectively, with transparent walls, open ends and variation of the cross-sectional dimensions have been conducted (see Fig. 11).

Fig. 12 shows a sequence of frames taken from high speed videography of this 1.8 meter long and 2 cm high (and deep) transparent duct. Each frame is arranged below the other. Thus, the vertical coordinate coincides with the time coordinate. The illuminated zone indicates the flame front spreading symmetrically to the left and right from the central point of ignition. The angle enclosed by this flame front line and the central vertical symmetry line indicates the flame speed. So it is easy to see the after a short deceleration phase after ignition the relatively pale blue flame is a relatively slow, but stable deflagration, which accelerates only after a certain time, to finally turn into a detonation. This deflagration-to-detonation-transition (DDT) is clearly marked by the strong flashing light emitted in the lower left and right corner of Fig. 12.

Fig. 11: Experimental set-up for transient flame behavior in micro-channels (red rectangle marks the cut-out of the transparent channel as used as a frame in Fig. 12

In total about 600 experiments have been conducted with variations of the concentration of H_2-O_2, location of the ignition point (in the middle or at the open end), and the cross-sectional area of the channel (quadratic with edge lengths of 4, 8, 10 and 20 mm). The influence of turbulence was investigated by applying a turbulence generating grid or spiral. Additionally the influence of confinement was tested for the open end ignition cases, by closing the channel end opposite to the ignition location. The main objective was to provide an overview on the transient behavior of the flame, manifested in the flame acceleration and deflagration-to-detonation transition. Additionally the detonation cell size was measured, linking the work to the assessment of more fundamental properties of these pre-mixed systems. The measured values have been compared to theoretical prediction by Gavrikov et al.

For both cases, with ignition in the middle and at the free end, a clear tendency for slower flames and a less probable DDT was observed for increasing channel cross-section dimensions. However, for the cases with ignition at the free end this was much more pronounced. No DDT was observed any more for channel larger than 8mm.

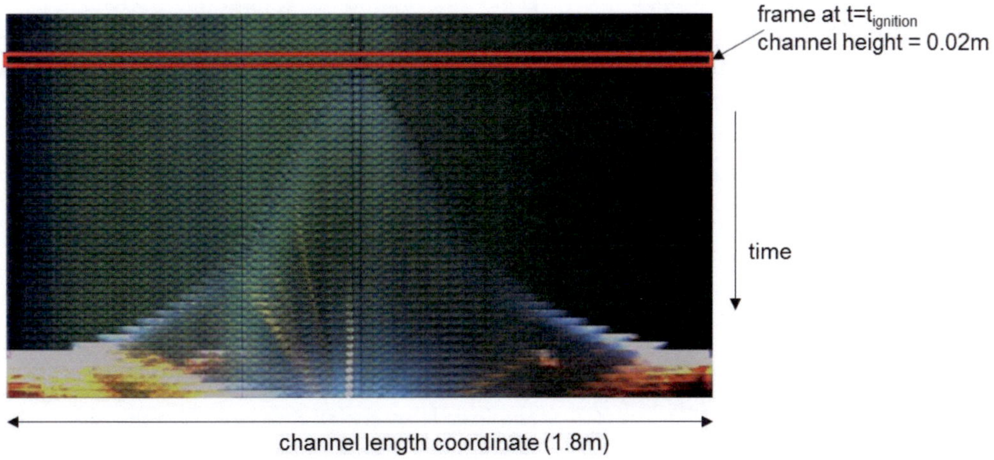

Fig. 12: Symmetric flame front progression and deflagration-to-detonation in transparent micro-channel

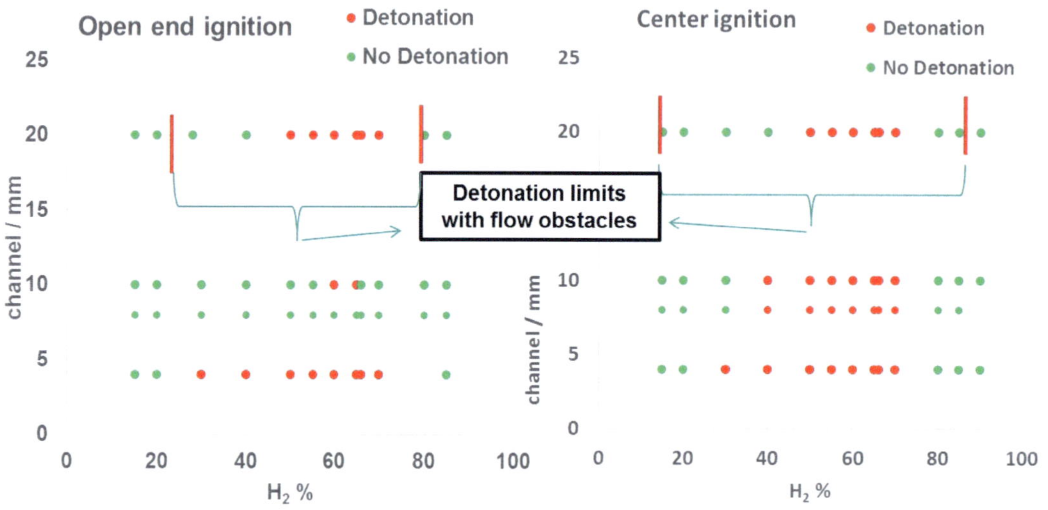

Fig. 13: Detonation limits for open end ignition (left) and center ignition for different concentration of H2 in O2 and for different channel dimension.

The ignition in the middle promoted much faster flames and shorter run-up distances to the DDT point. The concentration bandwidth, for which fast flames and DDT could be observed, is increasing with decreasing channel dimensions.

This DDT was strongly promoted in obstructed channels, what was realized in the experiment presented in Fig. 12 with a lengthy spiral introduced into the channel. This flow obstacle is visible in the figure as the dim horizontal zigzag lines, identically repeated in each frame.

In general, the stability of flame, effective flame speed and associated pressure developing in such narrow channels were found to be strongly dependent on the cross-sectional area, the congestion and mixing properties. The data generated with these experiments is summarised in Fig. 13 and is going to be used for further model and code validation.

References

[1] Paladino, D.; Andreani, M.; Guentay, S.; Mignot, G.; Kapulla, R.; Paranjape, S.; Sharabi, M.; Kisselev, A.; Yudina, T.; Filippov, A.; Kamnev, M.; Khizbullin, A.; Tyurikov, O.; Liang, Z. (R.); Abdo, D.; Brinster, J.; Dabbene, F.; Kelm, S.; Klauck, M.; Götz, L.; Gehr, R.; Malet, J.; Bentaib, A.; Bleyer, A.; Lemaitre, P.; Porcheron, E.; Benz, S.; Jordan, T.; Xu, Z.; Boyd, C.; Siccama, A.; Visser, D. (2016). Outcomes from the EURATOM-ROSATOM ERCOSAM SAMARA projects on containment thermal-hydraulics for severe accident management. Nuclear engineering and design, 308, 103-114. doi:10.1016/j.nucengdes.2016.08.011

[2] Ren, K.; Kotchourko, A.; Lelyakin, A. (2016). Adaptive LMR simulation of DDT in obstructed channel. 24th International Conference on Nuclear Engineering, ICONE 2016; Charlotte; United States; 26 June 2016 through 30 June 2016. Volume 4, Art. Nr: V004T10A031, ASME, New York (NY). doi: 10.1115/ICONE24-61105

[3] Royl, P.; Xiao, J.; Jordan, T. (2016). Blind simulations of THAI test TH27 with GASFLOW-MPI for participation in the international benchmark conducted within the German THAI program. OECD/NEA & IAEA Workshop Application of CFD/CFMD Codes to Nuclear Reactor Safety and Design and their Experimental Validation, Cambridge, MA, September 13-16, 2016.

[4] Royl, P.; Xiao, J.; Jordan, T. (2016). Double blind simulation of THAI test TH27 with GASFLOW-MPI for participation in the OECD benchmark defined and organized by Becker Technologies. OECD/NEA & IAEA Workshop Application of CFD/CFMD Codes to Nuclear Reactor Safety and Design and their Experimental Validation, Cambridge, MA, September 13-16, 2016.

[5] Xiao, J.; Kuznetsov, M.; Travis, J. (2016). Numerical modeling of heat losses from hydrogen jet flame in a vented combustion chamber. 8th International Seminar on Fire and Explosion Hazards (ISFEH), Hefei, China, April 25-28, 2016.

[6] Xiao, J.; Travis, J.; Bottoni, M. (2016). Thermal-hydraulic analysis of a passive containment cooling system using dynamic liquid film model in CFD code GASFLOW-MPI. 24th International Conference on Nuclear Engineering (ICONE24), Charlotte, NC, June 26-30, 2016. Proceedings on CD-ROM, ICONE24-60018, ASME, New York (NY).

[7] Xiao, J.; Travis, J.; Royl, P.; Necker, G.; Svishchev, A.; Jordan, T. (2016). GASFLOW-MPI: A Scalable Computational Fluid Dynamics Code for Gases, Aerosols and Combustion. Band 1 (Theory and Computational Model (Revision 1.0). (KIT Scientific Reports ; 7710). KIT Scientific Publishing, Karlsruhe. doi:10.5445/KSP/1000050393

[8] Xiao, J.; Travis, J.; Royl, P.; Necker, G.; Svishchev, A.; Jordan, T. (2016). GASFLOW-MPI: A Scalable Computational Fluid Dynamics Code for Gases, Aerosols and Combustion. Band 2 (Users' Manual (Revision 1.0). (KIT Scientific Reports ; 7711). KIT Scientific Publishing, Karlsruhe. doi:10.5445/KSP/1000050394

[9] Xiao, J.; Travis, J. R.; Bottoni, M. (2016). Validation of liquid-film transport model in a CFD code for passive containment cooling system using exact numerical solutions. 11th International Topical Meeting on Nuclear Thermal-Hydraulics, Operation and Safety (NUTHOS-11), Gyeongju, Korea, October 9-13, 2016. Proceedings on USB-Stick, Paper No. N11A0024.

[10] Xiao, J.; Travis, J. R.; Royl, P.; Necker, G.; Svishchev, A.; Jordan, T. (2016). Three-dimensional all-speed CFD code for safety analysis of nuclear reactor containment: Status of GASFLOW parallelization, model development, validation and application. Nuclear Engineering and Design, 301, 290-310. doi: 10.1016/j.nucengdes.2015.12.033

[11] Xiao, J.; Travis, J.; Bottoni, M. (2016). Generalized draining film model (DFM) equation set to execute an exact numerical solution. 12th International Conference on Heat Transfer, Fluid Mechanics and Thermodynamics, July 11-13, 2016, Malaga, E.

[12] Xu, Z.; Jordan, T.; Breitung, W. (2016). Particle Resuspension Model for Subatmospheric Conditions. IEEE transactions on plasma science, 44 (9), 1662–1665. doi: 10.1109/TPS.2016.2571120

[13] Xu, Z.; Jordan, T. L.; Breitung, W. (2016). Development of a particle resuspension model for a rarefied gas flow. 26th IEEE Symposium on Fusion Engineering, SOFE 2015, Austin, United States, 31 May 2015 through 4 June 2015, Art.Nr.: 7482296, IEEE, Piscataway, NJ. doi:10.1109/SOFE.2015.7482296

[14] Yanez, J.; Kuznetsov, M. (2016). An analysis of flame instabilities for hydrogen–air mixtures based on Sivashinsky equation. Physics Letters A, 380 (33), 2549-2560. doi: 10.1016/j.physleta.2016.05.048

[15] Yanez, J.; Kuznetsov, M. (2016). Experimental study and theoretical analysis of a 'strange wave'. Combustion and flame, 167, 494-496. doi:10.1016/j.combustflame.2016.02.004

[16] Zhang, H.; Xiao, J.; Travis, J. R. (2016). Direct numerical simulation and large eddy

simulation of a flow at low reynolds number over backward-facing step using the CFD code GASFLOW-MPI. 11th International Topical Meeting on Nuclear Thermal-Hydraulics, Operation and Safety (NUTHOS-11), Gyeongju, Korea, October 9-13, 2016. Proceedings on USB-Stick, Paper No. N11A0394.

Group: Areva Nuclear Professional School

Numerical Simulation of Proton Beam Induced Pressure waves in Spallation Targets

Jana Ruth Fetzer, Andreas Class

Introduction

In short pulsed accelerator driven neutron sources such as the Spallation Neutron Source (SNS) in Oakridge, TN and the Japanese Spallation Neutron Source (JSNS) heavy liquid metal is used as target material. The high energy proton beam deposits energy into the liquid during a sub-micosecond time frame [1]. This energy deposition results in very high heating rates. Since the timescale of the proton pulse is too short to allow the target material to expand during the pulse, the thermal expansion is impeded thus compressing the target material. This in turn results in the production of large amplitude pressure waves propagating through the target [2]. Once this pressure wave hits the target wall a negative pressure is induced due to the elasticity of the target wall. Moreover, if the short term tensile strength of the spallation material is met caviation in form of cavitation bubbles will occur. Collapse of cavitation bubbles in the vicinity of the target wall is accompanied by the characteristic damage pitting.

In case of a long pulsed spallation neutron source the amplitude of the induced pressure wave due to the proton beam heating is not as pronounced as for short pulsed sources. This is due to the lower energy per single pulse corresponding to the longer pulse length. The increased timescale allows the target material to partially adjust to the thermal expansion. Therefore, for a long pulse target the maximum pressure amplitude is no longer proportional to the total energy per pulse [3]. Instead, the sudden change in heating rate at the beginning and end of each pulse becomes the source for pressure waves. Yet all effects observed in short pulsed spallation targets may be relevant to long pulse targets as well. This includes induced pressure waves, cavitation and potentially pitting damage fortunately occurring at a lower repetition rate. The target container suffers during normal operation and beam trips substantial straining due the notable thermal volumetric expansion of the spallation material.

Even in continuous spallation sources using liquid metal cooled targets beam trips can induce pressure waves. A nuclear reactor application of a continuous source is proposed for the accelerator driven system MYRRHA (Multipurpose hYbrid Research Reactor for High-tech Applications). When the proton beam restarts after a beam interruption an initial high pressure zone is formed within the spallation zone. This high pressure zone then propagates outwards towards the target enclosure where reflection results in a negative pressure zone. If the tensile strength of the fluid is exceeded, cavitation and potentially pitting damage will occur.

The simulation of proton beam induced pressure waves are numerically challenging. Since applicable numerical time-steps become extremely small, the transient phenomena of pressure wave propagation need an exceedingly large number of time-steps. According to the CFL criteria the time-step is determined by the ratio of minimum grid size and characteristic flow velocity. In simulations covering pressure wave propagation compressibility of the fluid must be considered. Instead of the fluid velocity sound speed enters the CFL criteria. Unfortunately, the ratio of fluid velocity and sound speed, i.e. the Mach number is very small in target applications so that a pressure resolving simulation requires several orders more time-steps compared to an incompressible simulation.

Aero acoustics faces exactly the same problem. The Multiple Pressure Variable (MPV) approach was developed in aero acoustics to efficiently simulate weakly compressible flows. The MPV approach is based on a single time scale multiple space scale asymptotic analysis derived for subsonic flow by an asymptotic series expansion in the Mach-number.

We develop a MPV method for liquid metal applications and apply it to two target geometries, specifically the META:LIC [4] and MYRRHA targets. For each target, simulations are conducted which allow estimation of proton beam induced pressure waves.

Multiple pressure variables method for liquid metal application

A detailed asymptotic analysis of the compressible Euler equations has been performed by Klainerman and Madja [4]. Klein [6] extended this analysis to a multiscale asymptotic expansion for all physical variables $\boldsymbol{w} = \rho, \boldsymbol{U}, p$ and $w = f(\hat{x}, \xi, t) = f(\hat{x}, M\hat{x}, t)$ in powers of the Mach number

$$\boldsymbol{w} = \boldsymbol{w}^{(0)} + M\,\boldsymbol{w}^{(1)} + M^2 \boldsymbol{w}^{(2)} + \cdots. \quad (1)$$

Here x is the local variable associated with convective phenomena, while $\xi = Mx$ is a large scale coordinate and is associated with acoustic wave propagation. This asymptotic expansion is then substituted into the non-dimensionalized governing equations. All terms with same exponent of the Mach number are collected and separately equated to obtain a series of asymptotic limit equations. The asymptotic analysis shows constant spatially leading order pressure $p^{(0)}$. However, $p^{(0)}$ varies in time and serves to satisfy the equation of state [10], therefore is called thermodynamic pressure. Furthermore, the analysis clarifies that the pressure term $p^{(1)}$ is a function of the spatial variable. It is associated with acoustic waves hence it is called acoustic pressure. The second order pressure $p^{(2)}$ depends on temporal and both spatial scales. It assures the compliance with the incompressible $\nabla \cdot \boldsymbol{U} = 0$ contraint.

The characteristic Mach number of proton beam induced pressure waves in liquid metal spallation targets is due to the very low compressibility of liquid metals typically in the order of $\mathbf{10^{-4}}$. Due to these very low Mach numbers, an acoustic length scale becomes extremely long so distinction of the acoustic scale $\boldsymbol{\xi}$ is no longer applicable. Formally a numerical, yet more efficient two pressure variable approach suffixes, which was previously introduced by Park [8].

Governing equations

The compressible conservation equations for density $\boldsymbol{\rho}$, momentum $\boldsymbol{\rho U}$ and total energy per unit volume $\boldsymbol{E}$ are:

$$\partial_t \rho + \nabla \cdot (\rho \boldsymbol{U}) = 0 \quad (2)$$

$$\partial_t (\boldsymbol{U}\rho) + \nabla \cdot \big((\rho \boldsymbol{U}) \otimes \boldsymbol{U}\big) + \nabla p = \nabla \cdot \tau + \rho \boldsymbol{g} \quad (3)$$

$$\partial_t E + \nabla \cdot [\boldsymbol{U}(E + p)] = \nabla \cdot (\tau \boldsymbol{U}) - \nabla \cdot \boldsymbol{q} \quad (4)$$

where τ is stress tensor, p pressure, t time and g the gravitational acceleration.

The heat flux $\boldsymbol{q}$ is given by Fourier's law $\boldsymbol{q} = -\lambda \nabla T$, where the heat flux is associated with temperature T and the thermal conductivity λ.

Above system of equations is closed by the thermal and caloric equation of state of the fluid. The thermal equation of state for fluids is generally given in terms of pressure, temperature and specific volume. The caloric equation of state is provided in terms of internal energy or enthalpy [7]. To simulate general fluids, it is desirable to use a formulation of the energy equation in terms of pressure and enthalpy [9].

Neglecting without loss of generality thermal dissipation and hydrostatics the following equations are used to derive a multiple pressure variables formulation for liquid metal applications:

$$\partial_t \rho + \nabla \cdot (\rho \boldsymbol{U}) = 0 \tag{5}$$

$$\partial_t (\boldsymbol{U}\rho) + \nabla \cdot \left((\rho \boldsymbol{U}) \otimes \boldsymbol{U}\right) + \nabla p = \nabla \cdot \tau \tag{6}$$

$$\partial_t p + \boldsymbol{U} \cdot \nabla p + p\,(\nabla \cdot \boldsymbol{U}) = -\nabla \boldsymbol{q} - \rho \frac{\partial h}{\partial t} - \nabla \cdot (\rho\, \boldsymbol{U}\, h) \tag{7}$$

Equations (5) - (7) are non-dimensionalized so that the material properties appearing in the equations are replaced by dimensionless quantities. Therefore, this system of equations is valid for a class of physically similar problems. Reference length, velocity, density and pressure are used for non-dimensionalization. The specific enthalpy is made dimensionless with the reference velocity squared. Dimensionless independent and dependent variables are denominated $\hat{x}, \hat{t}, \widehat{\boldsymbol{U}}, \hat{\rho}, \hat{p}, \hat{h}$. The nabla operator in terms of non-dimensional spatial variables is $\widehat{\nabla}$. The equations read in non-dimensional notation:

$$\partial_{\hat{t}} \hat{\rho} + \widehat{\nabla} \cdot \left(\hat{\rho}\, \widehat{\boldsymbol{U}}\right) = 0 \tag{8}$$

$$\hat{\rho}\left(\partial_{\hat{t}} \widehat{U} + \left(\widehat{\boldsymbol{U}} \cdot \widehat{\nabla}\right)\widehat{\boldsymbol{U}}\right) = -M^{-2}\, \nabla \hat{p} + Re^{-1}\, \Delta \widehat{\boldsymbol{U}} \tag{9}$$

$$\partial_{\hat{t}} \hat{p} + \widehat{\boldsymbol{U}} \cdot \widehat{\nabla} \hat{p} + \hat{p}\left(\widehat{\nabla} \cdot \widehat{\boldsymbol{U}}\right) = -M^2 (\hat{\rho}\, \partial_{\hat{t}} \hat{h} + \widehat{\nabla} \cdot \left(\hat{\rho}\, \widehat{\boldsymbol{U}}\, \hat{h}\right) + (RePr)^{-1}\, \widehat{\nabla} \cdot \left(\lambda\, \nabla \widehat{T}\right)), \tag{10}$$

where $M = U_{ref}/c_{ref}$ represents the global Mach number with c_{ref} as the reference speed of sound. Reynolds and Prandtl numbers are denoted in suitable non-dimensional parameters Re and Pr, respectively.

The incompressible limit is as the Mach number tends to zero. The term $1/M^2$ multiplying the pressure gradient in the momentum equation shows singular behavior which is unfavorable for numerical methods. For simplicity hats denoting the dimensionless values are dropped below.

Multiple pressure variables approach

The reduced MPV approach only considers the leading and second order pressure terms. When eliminating the first order pressure or acoustic pressure its longwave contribution is contained within the second order pressure $p^{(2)}$. The reduced MPV ansatz introduced analogue to [8] is given by

$$p(x,t) = p^{(0)}(x,t) + M^2\, p^{(2)}(x,t) \tag{11}$$

Following the previous asymptotic analysis [4][6][8][10] the leading order pressure is defined as the average pressure within the whole computational domain Ω.

Inserting the reduced MPV Ansatz Equation (11) yields the following enthalpy equation:

$$M^2 \partial_t p^{(2)} + \partial_t p^{(0)} + (\boldsymbol{U}\,\cdot)\nabla p + p\,(\nabla \cdot \boldsymbol{U}) = -M^2((RePr)^{-1} \nabla \cdot (\lambda\, \nabla T) + \partial_t H + \nabla\,(\boldsymbol{U}H) \tag{12}$$

where $H = \rho \cdot h$ is enthalpy. This system of equations is implemented in our code.

Spallation Targets

META:LIC - MEgawattTArget:Lead bIsmuthCooled

The META:LIC target [4],[11] is a liquid metal target developed as comparative target solution for the European Spallation Source (ESS) according to ESS specifications [12]. Lead Bismuth Eutectic (LBE) is chosen as both target material and primary coolant. Fig 1 depicts the target module which is double walled. As shown, the target module consists of a proton beam guide with a safety window (not shown), an inflow channel leading to a nozzle producing a uniform block velocity profile, a U-bend with an expansion chamber and spoiler enforcing flow detachment to counteract the effects of the thermal expansion and pressure waves in the LBE due to the pulsed proton beam, and an outflow duct. The flow is pumped upwards into the inclined inflow channel, then accelerated by the nozzle. Next the LBE flows through the proton beam interaction zone which is inclined relative to the horizontal plane by 15 degrees. The LBE returns to the pool through an U-bend and outflow duct. A horizontal proton beam enters the liquid metal through a solid wall approximately 1.5mm thick. The small inclination angle of the proton beam interaction zone provides almost coaxial LBE flow and proton beam. This results in a fairly uniform heating of the coolant so that a minimal coolant flow rate can be established. Moreover, the flow component perpendicular to the proton beam transports the fluid across the beam in a fairly short time. This is advantageous for pulsed beams, as successive beam pulses interact with fluid that was not subjected to the beam previously.

The META:LIC target suffers of both, pressure pulses due to the pulsed nature of the proton beam and the strain on the target container due to the thermal expansion of the LBE [3].

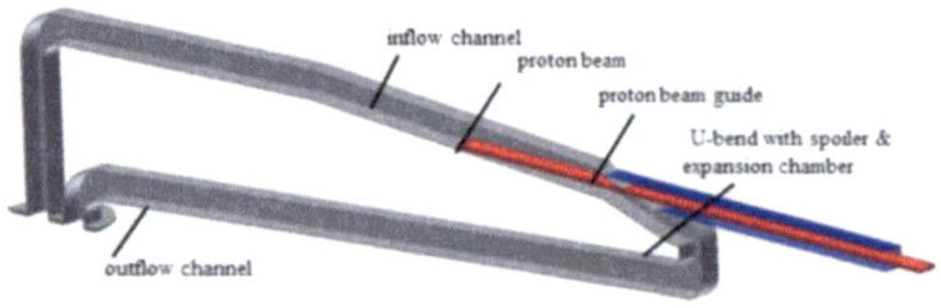

Fig. 1: META:LIC window target module

MYRRHA - Multi-purpose hYbrid Research Reactor for High-tech Applications

The MYRRHA reactor is a flexible experimental ADS system currently under development at SCK-CEN, demonstrating the coupling of an accelerator, spallation target and subcritical core at a reasonable power level. Furthermore, it is planned as a flexible irradiation facility and its reactor is supposed to work both in critical and subcritical mode [13], [14]. In an ADS, the reactor runs in subcritical mode. The spallation neutrons provided from the accelerator and target are needed to maintain fission. When MYRRHA is operated in ADS mode, the central position of the core is occupied by the so called spallation target assembly.

A sketch of the planned tube type spallation target is shown in Fig. 2. As depicted in the figure, the proton beam is guided by the beam tube into the central core region, where it enters the spallation zone through the proton beam window, which separates the accelerator vacuum from the target material. The proton beam window is cooled by the upward flowing primary coolant under forced convection supported by the primary pumps.

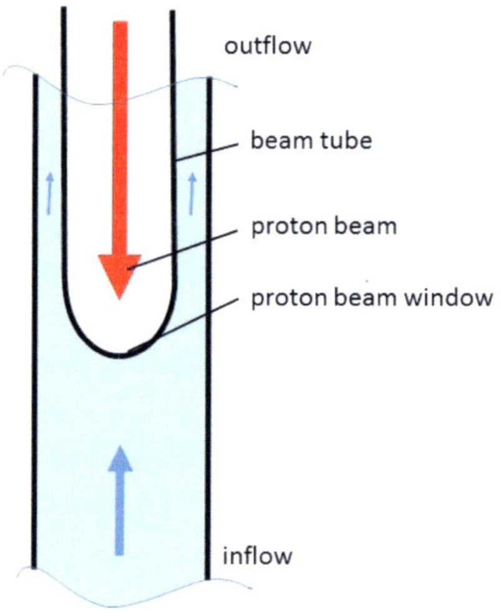

Fig.2: Sketch of the spallation region of MYRRHA window spallation target assembly.

META:LIC target calculations

Results of the MPV calculations for the META:LIC target of a single proton beam pulse are shown for various spatial and temporal resolutions. Initial conditions correspond to fully developed flow. The caloric and thermal equations of state for LBE proposed in [15] are used.

Geometry, boundary and initial condition

The geometry used for the META:LIC simulations as well as the applied boundary patches are displayed in Fig. 3. Due to symmetry considerations only half of the target width is simulated. The inflow velocity is *1.5 m/s* as specified. At the wall boundaries, a no slip condition is applied. At the inlet and outlet, wave transmissive boundary conditions reduce undesirable pressure wave reflections. The inlet temperature is set to *500 K*. Using the averaged speed of sound of LBE within the simulated geometry, a global Mach number of $M = 0.0015$ is determined. For the calculations, three different mesh resolutions were used, namely *0.37, 0.91* and *2.12* million cells. Five different time-steps are analysed: *2, 0.4, 0.2, 0.02* and *0.01 ms*. Five time-steps are used to find the optimal balance between accuracy, resolution and dissipation of the numerical scheme.

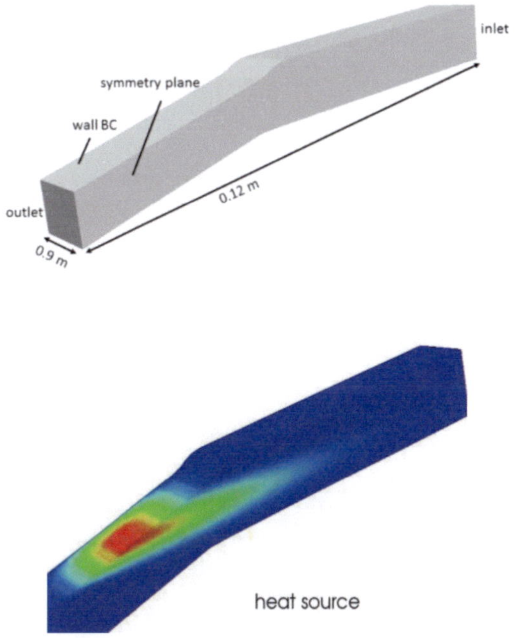

Fig. 3: Geometry and boundary conditions (left), power density distribution in [W/m³] (right) for the META:LIC target

The proton beam specifications are taken from [12] and displayed in Table 1. For all presented calculations, a single proton beam pulse is assumed. Fig. 3 (bottom) displays the power density distribution of the proton beam in the liquid metal target META:LIC.

Table 1: Proton beam specifications for ESS and MYRRHA.

	META:LIC	**MYRRHA**
Time structure	Pulsed	Continuous (microstructure)
		200 µs beam holes (at 1 — 250 Hz) (nominal macrostructure)
Pulse length	2 ms	
Pulse repetition rate	14 Hz	
Beam energy	2.5 GeV	600 GeV
Average beam current	2 mA	0.1 – 4 mA
Beam profile	Parabolic 160 mm x 60 mm	Circular Ø 85 mm

Simulation Results

The velocity and temperature distribution at the end of a single proton beam pulse is shown in Fig. 4. There is no visible distortion of the velocity field due to the thermal heating of the LBE. Fig. 4 (bottom) displays the total temperature rise of about $90\ K$ during a single proton beam pulse.

Snap shots of the pressure during a single proton pulse on the coarse mesh for $\Delta t = 0.02$ ms is displayed in Fig. 5. Fig. 5 (a) shows the pressure in the target module at the beginning of the pulse when the proton beam first hits the liquid metal. During this stage a pressure amplitude of approximately 7 bar is attained. The pressure then rises during the continued energy deposition up to approximately 17 bar at the end of the pulse, when the heat deposition stops, see Fig. 5 (b). Once the heat deposition stops, a negative pressure zone is formed due to the inertia of the surrounding fluid, which still conforms to the thermal expansion. Fig. 5 (c) depicts the pressure distribution approximately 2ms after the heat deposition stops. A negative pressure of *–10 bar* is reached. In the algorithm, strain is limited to *-10 bar*. In reality, the physical limitation is given by an unknown tensile strength of LBE. Exceeding the limit leads to cavitation.

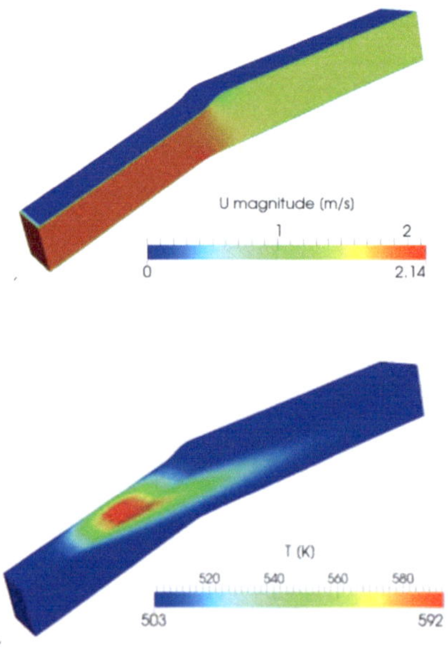

Fig. 4: Velocity magnitude in [m/s] (top) and Temperature in [K] (bottom) distribution at the end of a single proton pulse

Fig. 6 displays the temporal evolution of a single pressure pulse for different numerical time-steps on the mesh with *0.37* million cells. The largest chosen time-step corresponds to the proton beam pulse length, and the smallest is 200 times smaller. The blue line in Fig. 6 (top) displays the resulting pressure evolution if the pressure pulse is under resolved with a single time-step. The red dashed line shows very low temporal resolution with a ten times smaller time step. Comparing the maximum pressure amplitude of these two calculations Fig. 6 shows a difference of approximately *5 bar*, the same discrepancy can be observed for the negative pressure after the heat deposition stops. Using a time-step of *0.01 ms* results in a pressure amplitude of approximately *14 bar* and a negative pressure of *–7 bar*. From Fig. 6 it follows that the pressure peak is largest for a time step of 0.2 to 0.4 ms. Small time-steps increase numerical dissipation and thus the pressure amplitude. However, it can also be observed that increasing time resolution allows for fluctuations of the pressure during the pulse. These fluctuations are due to the now resolved pressure reflections at the wall boundaries. For conservative estimations, we propose to use intermediate time-steps.

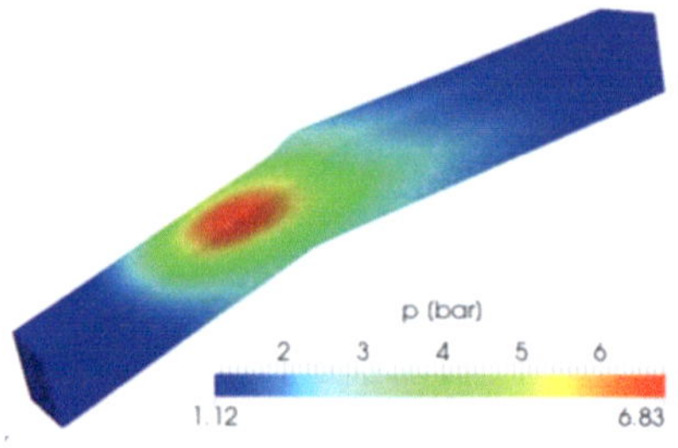

(a)

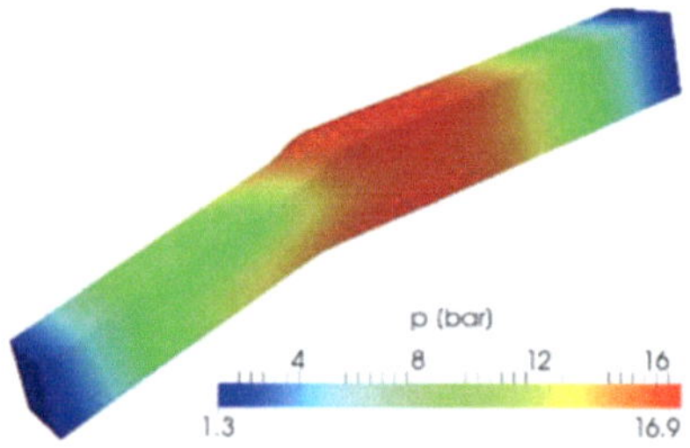

(b)

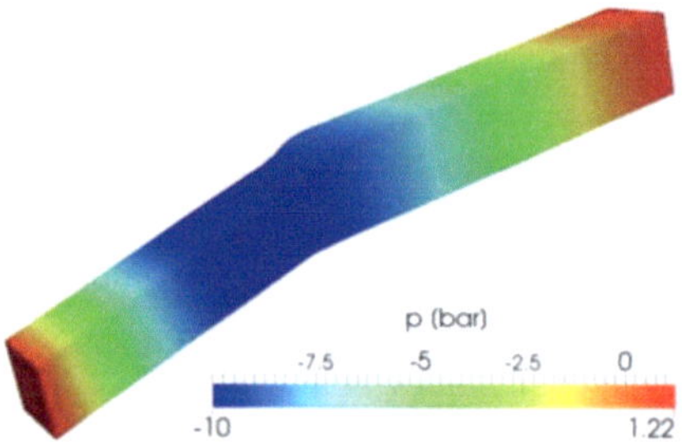

(c)

Fig. 5: Pressure distribution at beginning of pulse (a), end of pulse (b) and at t = 1.077s for a time-step of 0.02ms.

Fig. 6 (top) displays the influence of the mesh resolution on the pressure evolution. Here, the pressure as a function of time is shown for three different numerical grids and two different time-step length. The figure shows that there is virtually no difference in the pressure evolution for the different meshes. This is the case for both time-steps

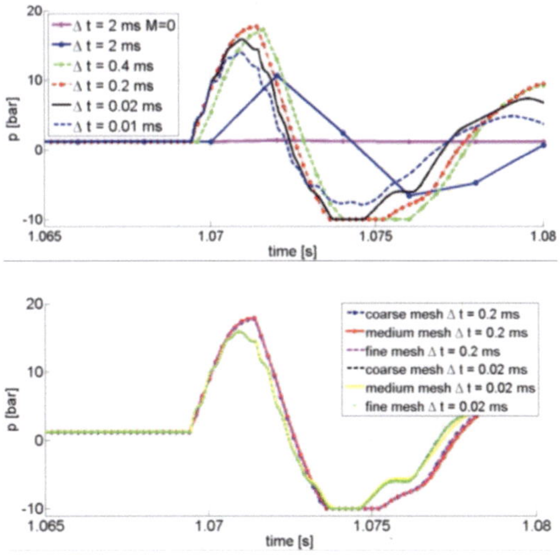

Fig. 6: Pressure [bar] vs time [s] for different time-steps (top) and pressure [bar] vs time [s] for different spatial resolutions (bottom).

MYRRHA target calculations

Geometry, boundary and initial conditions

A simplified axisymmetric two-dimensional geometry is used for the simulations. The influence of the original hexagonal shape of the assembly on the resulting fluid flow is neglected for a first investigation of proton beam induced pressure waves. Fixtures, such as flow straighteners downstream of the proton beam interaction zone, are disregarded for the same reason. Fig. 7 (top) displays the simulated geometry as well as the applied boundary patches.

At the inlet boundary condition, an inflow velocity of *0.45 m/s* is chosen to limit the outlet velocity in accordance to the erosion limit. At the wall boundaries, a no slip condition is applied. At the inlet and outlet, wave transmissive boundary conditions reduce undesirable pressure wave reflections at both. The inlet temperature is set to *543 K*.

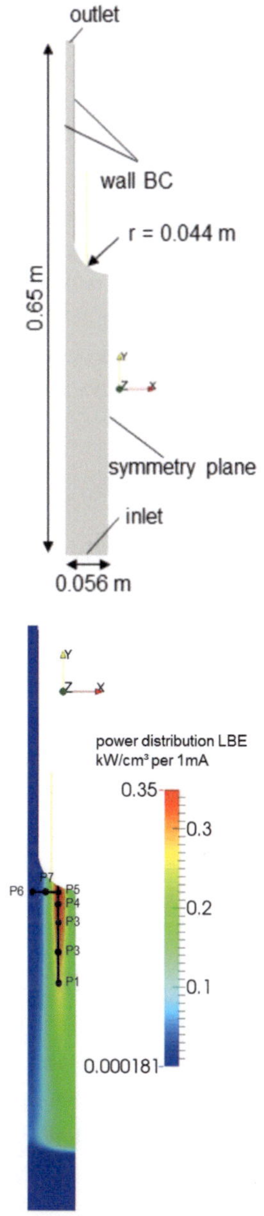

Fig. 7: MYRRHA geometry and boundary conditions (top), power deposition and sample points (bottom).

For the calculations, the global Mach number is *0.0005*. The behavior of the pressure is evaluated at several sample points (P1 –P5), which are located along a straight line in the area with the highest energy deposition. P5 is in the region with the highest energy deposition and also closest to the proton beam window, point

P1 is 0.155m upstream of P5. Fig. 7 (top) displays the position of the sample points.

The specifications of the MYRRHA accelerator are taken from [16] and listed in Table 1. For the CFD simulations, a peak current of 4mA is assumed. According to [16], the nominal time macrostructure of the proton beam features 200 µs beam holes at frequencies of 1 – 250 Hz. Simulated frequencies are 100Hz and 250 Hz. The intended sweeping of the proton beam is not considered in present calculations. This sweeping might result in slightly higher pressure wave amplitudes. Fig. 7 (bottom) displays the power distribution in LBE in kW/cm³ per 1 mA.

Nominal proton beam operation at 250Hz

For this simulation, a nominal macrostructure with 200 µs beam holes at a frequency of 250 Hz is assumed. Fig. 8 displays the pressure history at sample point P5 (red dashed line) and P1 (blue continuous line) for a simulation time of 0.05s, which corresponds to 13 beam pulses. The figure shows that for normal operation conditions the pressure fluctuates by approximately ±0.2 bar at Point P5, the area with the highest energy deposition and ±0.15 bar at Point P1.

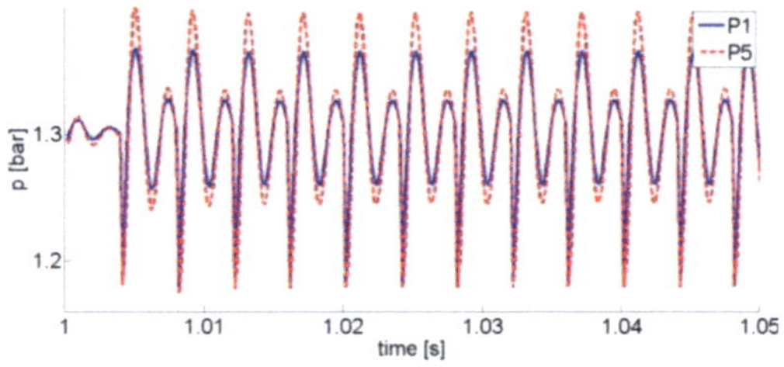

Fig. 8: Pressure [bar] vs time [s] with nominal beam operation at 250 Hz

Beam trip results for 100 Hz and 250 Hz

The investigation of the influence of beam trip on the temporal behavior of the pressure at Point P1 has been performed for a nominal macrostructure frequency of 100Hz and 250Hz. The simulated beam trips begin with normal proton beam operation as shown in Fig. 8. Then the beam is interrupted for a predefined period. In the simulated case this was a beam trip of 0.01 and 0.001s. These short beam trips are not limited in occurrence according to the MYRRHA accelerator beam requirements given in [16]. Following a beam trip, the proton beam returns to normal operation. The time-step width for the simulation is chosen to be $1*10^{-4}$ s. Therefore, the beam trip with a duration of 0.001 s is discretized by 10 time-steps and the second beam trip with 100 time-steps.

Fig. 9 depicts both described beam trips for a nominal operation macrostructure of 100Hz (left) and for an operation macrostructure of 250Hz (right). For the shorter beam trip duration, depicted by the blue continuous line, peak pressure amplitudes of approximately 2 bar and 2.1 bar are calculated. For the second beam trip, with a duration of 0.01s shown by the red dashed line, the peak pressures reach about 1.75 bar. For both macrostructures and beam trip durations, the maximum pressure drop due to the sudden loss of proton beam heating is about 0.45 bar. The pressure oscillations following the peak pressure values are mainly due to wall boundary reflections. Small inlet outlet reflections are still observable and could be omitted by extending the computational domain. For an experimental facility, these pressure pulses may be tolerable, but for long term ADS operation fatigue should be addressed.

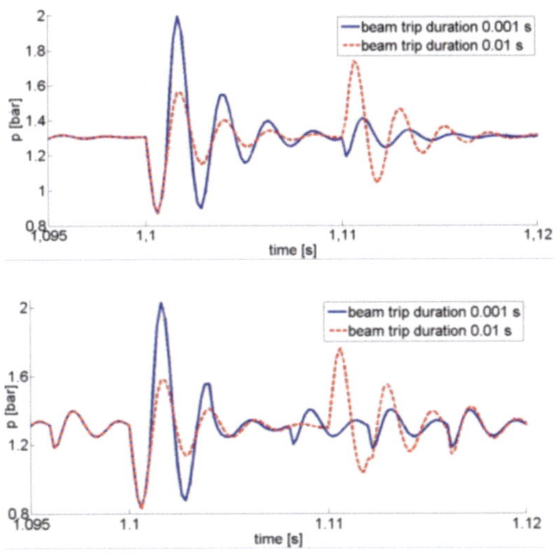

Fig. 9: Pressure [bar] vs time [s] for beam trip durations of 0.01s and 0.001s for 100 Hz (top) and 250 Hz (bottom).

Conclusion

Liquid metal spallation targets are subjected to time dependent heat deposition for both continuous and pulsed operation. With a traditional CFD approach, analysis of beam trips and the pulsed temporal structure require extensive numerical simulations. In particular, the acoustics limit forces extremely short temporal time steps. The simulation of the MYRRHA and META:LIC heavy liquid metal target using MPV method demonstrate that the restrictive time-step limitation can be overcome by the asymptotic approach. Both targets show pressure pulses with an amplitude of the order of bars which need consideration since pitting damage or fatigue can occur and should be controlled. Final verification of this damage requires additional experimental investigations on the tensile strength of the chosen liquid metal.

References

[1] T. McManamy et al. (2008) J. Nucl. Mater. 377, pp. 1–11

[2] T. A. Gabriel, et al. J (2001*)* Proc. 2001 Particle Accelerator Conference, 2001.

[3] J.R. Fetzer, A.G. Class, J. Wolters (2014) J Nucl. Mater. 450

[4] A.G. Class, C. Fazio, J.R. Fetzer, S. Gordeev (2014) J Nucl. Mater. 450

[5] S. Klainerman, A. Majda (1982) *Comm: on Pure and Appl. Math.*, 35

[6] R. Klein (1995*) J.* Comp. Phys., 121

[7] Stephan, K. et al. (2013) *"Thermodynamik Grundlagen und technische Anwendungen Band1: Einstoffsysteme"* Springer ISBN 978-3-642-30098-1

[8] J.H. Park, C.-D. Munz (2005*)* J. for Numerical Methods in Fluids, 49,

[9] L.M.T. Sommers (1994) PhD Thesis TU Eindhoven

[10] C.-D. Munz, S. Roller, R. Klein, K.J. Geratz (2003) *"The extension of incompressible flow solvers to the weakly compressible regime"*, Computer & Fluids, 32, pp. 173-196

[11] B. Weinhorst, U. Fischer, A.G. Class, J.R. Fetzer (2014) *"Radiation damage analysis for the liquid metal target META:LIC"* J Nucl. Mater. 450, pp. 219-224

[12] S. Peggs (2013) *"ESS Technical Report"* http://eval.esss.lu.se/DocDB/0002/000274/014/TDR_online_ver_all.pdf

[13] H.A. Abderrahim et a.. (2012*)* Energy Conversion and Management 63

[14] J. Engelen, et al. (2015) Int. J. Hydrogen Energy 40

[15] V.P. Sobolevet et al. (2008), J. Nucl. Mater. 376

[16] D. Vanderplassche et al. (2014) "*MAX Project Final Report"*

Read more:

[17] Benz, M. (2016). Statistische Modellierung einer geschichteten Zweiphasenströmung (KIT Scientific Reports ; 7705). Dissertation. KIT Scientific Publishing, Karlsruhe. doi:10.5445/KSP/1000056168

[18] Hartig, M.; Schulenberg, T.; Batta, A. (2016). Numerical study of fluid-structure interaction with heat transfer at supercritical pressure in a fuel rod assembly. Journal of Nuclear Engineering and Radiation Science, 2, 011013/1-12. doi:10.1115/1.4031816

[19] Noack, R. (2016). Multiskalen-Schließung in der Grobgitter-Methode zur Simulation zweiphasiger Strömungen (KIT Scientific Reports ; 7706). Dissertation. KIT Scientific Publishing, Karlsruhe. doi:10.5445/KSP/1000052392

[20] Raque, M.; Schulenberg, T.; Zeiger, T. (2016). Expected safety performance of the supercritical water reactor fuel qualification test. Journal of Nuclear Engineering and Radiation Science, 2, 011005/1-9. doi:10.1115/1.4030917

[21] Roelofs, F.; Shams, A.; Batta, A.; Moreau, V.; Di Piazza, I.; Gerschenfeld, A.; Planquart, P.; Tarantino, M. (2016). Liquid metal thermal hydraulics, state of the art and beyond: The SESAME project. Proceedings of the European Nuclear Conference (ENC 2016), Warszawa, PL, October 9-13, 2016, 77-88, European Nuclear Society, Brussels.

[22] Rohde, M.; Peeters, J. W. R.; Pucciarelli, A.; Kiss, A.; Rao, Y. F.; Onder, E. N.; Muehlbauer, P.; Batta, A.; Hartig, M.; Chatoorgoon, V.; Thiele, R.; Chang, D.; Tavoularis, S.; Novog, D.; McClure, D.; Gradecka, M.; Takase, K. (2016). A blind, numerical benchmark study on supercritical water heat transfer experiments in a 7-rod bundle. Journal of Nuclear Engineering and Radiation Science, 2 (2), 021012. doi:10.1115/1.4031949

[23] Ruzickova, M.; Vojacek, A.; Schulenberg, T.; Visser, D. C.; Novotny, R.; Kiss, A.; Maraczy, C.; Toivonen, A. (2016). European Project 'Supercritical Water Reactor-Fuel Qualification Test': Overview, Results, Lessons Learned, and Future Outlook. Journal of Nuclear Engineering and Radiation Science, 2, 011002/1-10. doi:10.1115/1.4031034

[24] Schulenberg, T. (2016). Status and future directions of SCWR development. Utilization of SUSEN (SUStainable ENergy) Technologies for the Gen IV Development, Husinec-Řež, CZ, November 24, 2016.

[25] Schulenberg, T. (2016). Alternativen zur nuklearen Entsorgung. Landesfachausschuss Forschung, Technologie und Energie, Frankfurt, 15.Oktober 2016.

[26] Schulenberg, T. (2016). Guest Editorial - Special Issue: Seventh International Symposium on Supercritical Water-Cooled Reactors. Journal of Nuclear Engineering and Radiation Science, 2, 010301/1-2. doi:10.1115/1.4031661

[27] Schulenberg, T.; Leung, L. (2016). Super-critical water-cooled reactors. Handbook of Generation IV Nuclear Reactors. Ed.: I.L. Pioro, 189-220, Elsevier, Amsterdam. doi:10.1016/B978-0-08-100149-3.00008-2

[28] Schulenberg, T.; Sierraalta Casas, I. A.. Simulation of transient heat transfer in an SCWR fuel assembly test at near-critical pressure. IAEA Technical Meeting on Heat Transfer, Thermal-Hydraulics and System Design for SCRWs, Sheffield, GB, August 22-24, 2016.

[29] Starflinger, J.; Schulenberg, T. (2016). The mutual influence of thermal-hydraulics and materials on design of SCWR. Review of results of the project HPLWR phase 2. IAEA Technical Meeting on Heat Transfer, Thermal-Hydraulics and System Design for SCRWs, Sheffield, GB, August 22-24, 2016.

[30] Viellieber, M. (2016). Coarse-Grid-CFD für industrielle Anwendungen: Integrale Analysen detaillierter generischer Simulationen zur Schließung von Feinstrukturtermen eines Multiskalenansatzes (KIT Scientific Reports ;

7727). Dissertation. KIT Scientific Publishing, Karlsruhe. doi:10.5445/KSP/1000062943

[31] Vojacek, A.; Ruzickova, M.; Schulenberg, T. (2016). Design of an in-pile SCWR fuel qualification loop. Journal of Nuclear Engineering and Radiation Science, 2, 011003/1-7. doi:10.1115/1.4030872

Scaling and Outgassing in Geothermal Power Plants – Influencing Factors, Kinetics, Inhibitors and Technical Measures

Tina Canic, Dietmar Kuhn, Sabine Baur, Dominik Mayer

In 2016, a four-year project founded by the German Federal Ministry for Economic Affairs and Energy has been successfully completed. The project dealt with the serious issues of scaling and outgassing in geothermal power plants in Germany. Several influencing parameters have been studied and many aspects of the scaling and outgassing problems have been addressed. When knowing how different parameters influence these effects, plant operators can handle them better. This will improve system efficiency and running costs.

Barite scaling

In geothermal power plants, geothermal brines from several kilometres below the earth surface are used for power generation. Depending on local conditions they can contain up to 350 g/L of dissolved salts [1]. In the case of pressure and temperature changes, these salts can become supersaturated and thus precipitate.

Especially in the Upper Rhine Graben, barium sulfate (barite, $BaSO_4$) is a serious issue. It is an extremely insoluble salt [2, 3], the petroleum industry speaks of a "particularly intractable scaling agent" [4]. Once formed, it cannot be removed by the addition of acids; in fact, it requires mechanical removal.

The precipitation of barite from a solution with barium and sulfate ions is influenced by different parameters. These include

- Supersaturation
- Overall salt concentration
- Shear rate (flow conditions)
- Contact surface
- Crystallization nuclei
- Molar ratio $n(Ba^{2+})/n(SO_4^{2-})$

For the experiments with constant shear rates, an annular gap apparatus was used (Fig. 1). Details of the experimental setup, the procedure and the measurement of the reaction progress are given in [5, 6]. The dependency of the barite precipitation rate on the other parameters was also studied in stirred solutions in glass vessels.

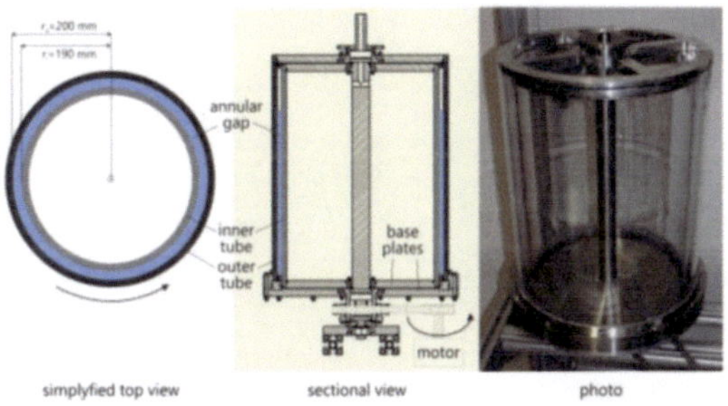

Fig. 1: Annular gap apparatus

The starting point of each reaction experiment was the separate addition of barium and sulfate ions to the following model brines:

- Type North German Basin (NGB): 104 g/L NaCl and 161 g/L $CaCl_2$
- Type Upper Rhine Graben, I (URG-I) 77.1 g/L NaCl and 24.8 g/L $CaCl_2$
- Type Upper Rhine Graben, II (URG-II) 70.6 g/L NaCl and 21.9 g/L $CaCl_2$

Fig. 2 shows the barium ion concentration versus time in two different model brines (NGB: violet and pink, URG-I: dark and light green) at two different supersaturations (given by the saturation index *SI*, higher values mean higher supersaturation).

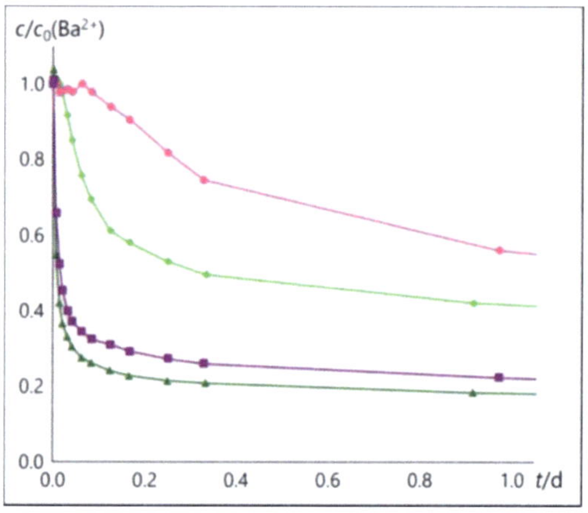

Fig. 2: Barium ion concentration *c* normalized to the initial concentration c_0 versus time *t* at different saturation indices *SI*

♦	NGB	SI = 1.47
♦	URG-I	SI = 1.46
■	NGB	SI = 1.96
▲	URG-I	SI = 1.96

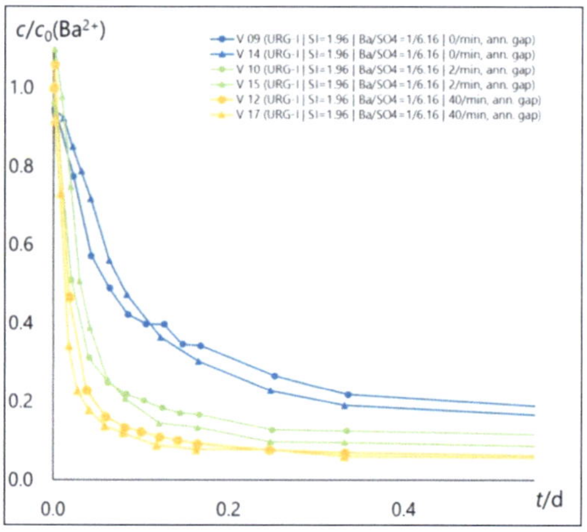

Fig. 3: Barium ion concentration *c* normalized to the initial concentration c_0 versus time *t* at different shear rates (*SI* = 1,96):

Blue: two identical experiments at 0 min^{-1}

Green: two identical experiments at 2 min^{-1}

Yellow: two identical experiments at 40 min^{-1}

A decreasing barium ion concentration means that the concentration of precipitated barite is rising at the same time. At both supersaturations, the barium ion concentration decreases faster in the URG-I brine (green lines) with the lower overall salt concentration. This effect is more pronounced at the lower supersaturation. Comparing the two supersaturations, barite precipitates slower at the lower supersaturation in both model brines.

Fig. 3 shows the barium ion concentration versus time at three different rotational speeds of the annular gap apparatus, thus representing different shear rates. Lines of the same color represent two experiments at identical conditions. It can be seen that with higher shear rate the precipitation becomes faster.

From all the experiments, the following dependencies result: The barite precipitation rate rises with

- lower overall salt concentration
- rising supersaturation
- higher shear rate
- when providing crystallization nuclei
- with higher sulfate excess

Targeted crystallisation

Crystallization nuclei accelerate the barite precipitation from geothermal brines. This process has been studied in a crystallization reactor (Fig. 4). A $BaCl_2$ solution and a Na_2SO_4 solution, both based on a model brine, were mixed continuously in a y mixer to form barite. The mixture then passes the reactor, filled with a specific amount of barite crystallization nuclei. The Barite concentration was measured after passing through the reactor at different retention times.

After a retention time of 1 minute, an experiment run at 80 °C showed a decrease in barium ion concentration: without nuclei, the concentration decreased to 90 % of the initial concentration, with the nuclei, the concentration falls to 78 %.

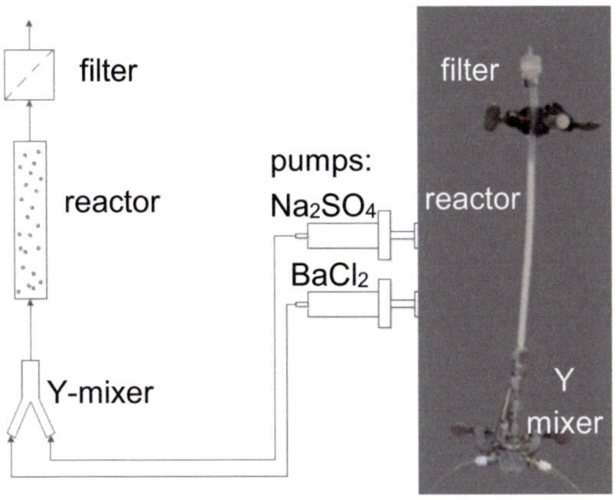

Fig. 4: Reactor for the targeted crystallization of barite on crystallization nuclei.

With this result, the construction of a crystallization apparatus for the targeted crystallization of barite would be a promising concept. With such an apparatus, a certain amount of barite crystals can precipitate at a specific location in the plant, thus lowering the supersaturation of barite and the probability that barite precipitates in critical plant components.

Geochemical monitoring in Unterhaching

Among other parameters, the gas content and composition of the brine have been monitored. The geothermal brine in Unterhaching in the German Molassebecken contains 200 standard milliliters of gas per liter fluid. This value showed a fluctuation of only 5 % during the project (monitoring once a year from 2012 to 2015). The composition of the gas will be shown in Fig. 7; these values have been measured between April and July 2014.

The amount of total dissolved salts has only slightly increased by 10 % from 614 mg/L (2012) to 676 mg/L (2015). This increase can be due to several influencing factors, e. g. cleaning procedures with acids.

Precipitation and degassing bypass

The precipitation and degassing bypass (Fig. 5) has been developed at the IKET and aims at the on-site investigation of precipitation from geothermal brines and the online analysis of gases at different pressures. It consists of a degassing unit, DB1, where the gas content and the gas composition can be determined via the liquid level and gas chromatography, respectively.

WT1, WT2, and WT3 are heat exchanger units kept at different lower but constant temperatures. They are used to determine the precipitation from the geothermal brine at different temperatures. The bypass is connected to the geothermal water loop of the plant via a heated transfer line.

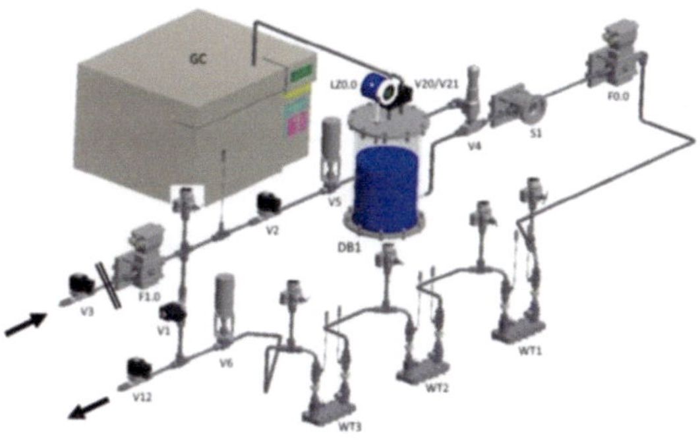

Fig. 5: Precipitation and degassing bypass apparatus

To determine the gas flow rate temperature, pressure and volume flow were kept constant, and the fluid level of the degassing unit was measured. From the rate of level change, the gas flow rate was calculated.

Fig. 6 shows the gas flow rate when degassing the geothermal fluid at different pressures. It can clearly be seen, that at a pressure higher than 4 bar no outgassing occurs. With decreasing pressure, the gas flow rate increases.

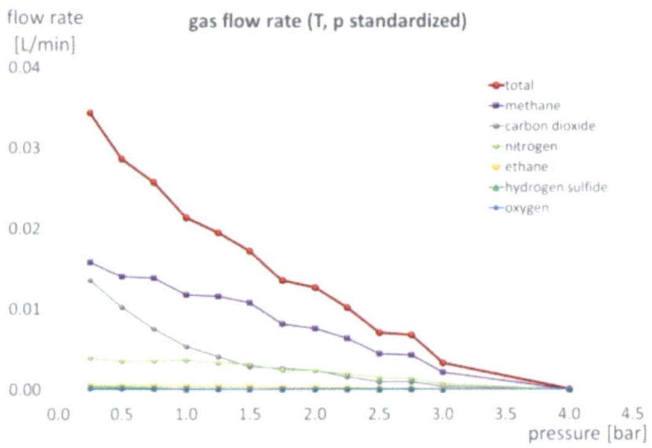

Fig. 6: Gas flow rate at different pressures

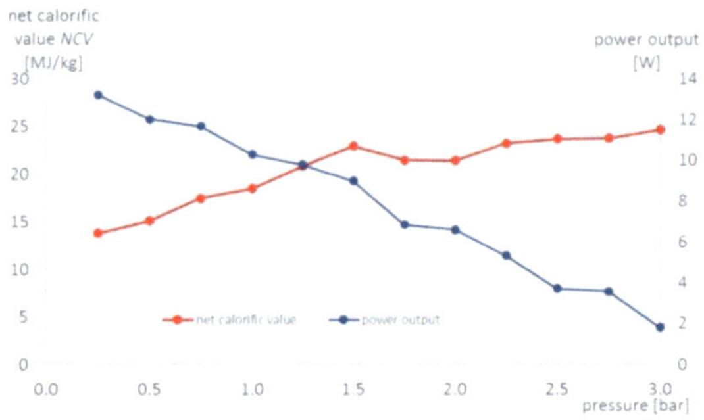

Fig. 8: Net calorific value *NCV* of the gas phase and power output.

Fig. 7 shows the composition of the gas at different pressures, as measured online by the gas chromatograph. At nearly all pressures, the methane ratio is above 50 % by volume. With rising pressure the CO_2 ratio falls whereas the methane and nitrogen ratios increase.

Fig. 8 shows the net calorific value of the gas and the resulting power output. The net calorific value of the gas is between 14 and 25 MJ/kg, which is in the same order of magnitude as town gas and thus the "geothermal gas" could be used for combustion.

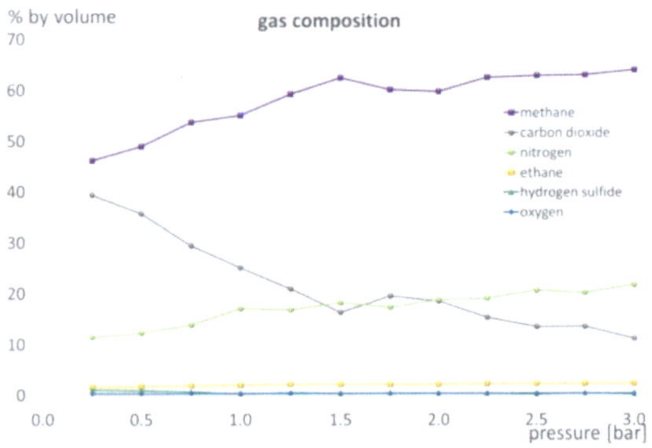

Fig. 7: Composition of the gas phase at different pressures.

New measurement and diagnosis tools

Outgassing of dissolved gases in geothermal brines leads to two-phase-flow in the power plant causing pressure shocks that can damage plant components. To avoid this, an ultrasonic measurement procedure has been developed helping to detect gas bubbles in pipe flows (Fig. 9).

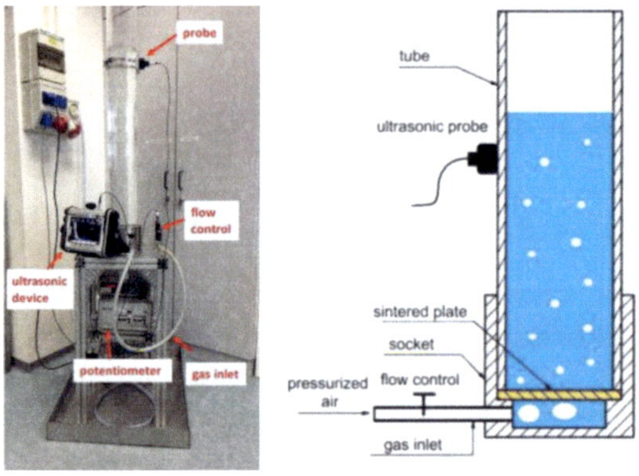

Fig. 9: Test facility for the ultrasonic apparatus measuring the void content

An ultrasonic signal is sent through a pipe, reflected at the opposite pipe wall and then detected. Fig. 10 shows a typical ultrasonic signal. The peaks on the left represent the reflections of the first inner wall of the tube; the peaks on the right represent the reflections of the back wall.

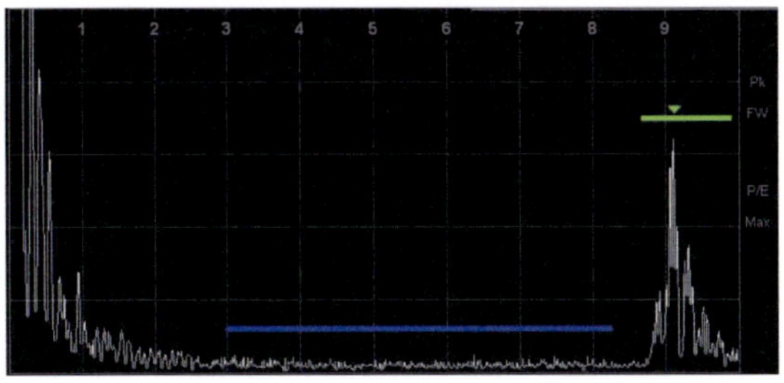

Fig. 10: Typical ultrasonic signal without bubbles

Bubbles present in the pipe will reduce the back wall echo (marked with the green line in Fig. 10) and increase the noise of the back wall echo. These parameters can be used to measure the amount of undissolved gas in the brine. A more sensitive parameter, that responds even to small amounts of gas bubbles, is the coefficient of variation of the back wall echo.

Fig. 11 shows the dependency of the coefficient of variation on the gas content at stainless steel tubes with different wall thicknesses. Even with the thickest tube, this parameter is very suitable to detect the smallest amounts of gas bubbles of about 0.2 %.

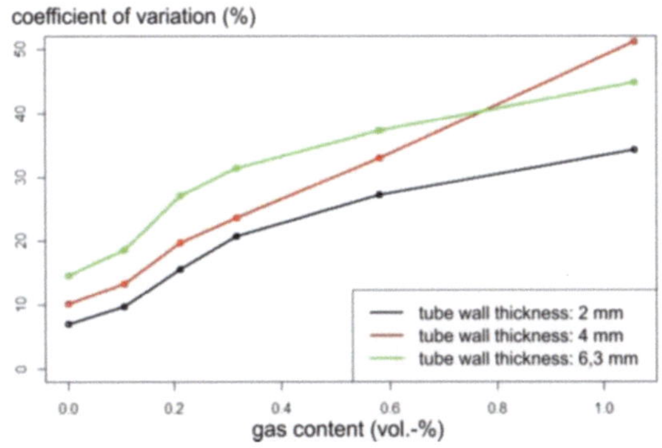

Fig. 11: Coefficient of variation of the back wall echo versus gas content in tubes with different wall thicknesses

Another tool, a quake catcher, has been developed to detect hydrostatic locks in the piping system of a geothermal plant. It consists of an acceleration sensor, a memory card and a battery connected to an Arduino Uno microcontroller board. The Arduino was programmed to record the accelerations in x-, y- and z-direction and to write them on the memory card. The quake catcher can easily be mounted on the piping system of the plant and show when and how many hydrolocks occur during a certain period of time.

Inhibition of calcium carbonate

In Unterhaching, calcium carbonate ($CaCO_3$) scaling is formed during plant operation. To prevent the formation of $CaCO_3$ crystals, different anorganic and organic inhibitors have been tested. Polyelectrolytes can form dative bonds with cations or anions and thus separate the ions from their counterparts. Those substances can block crystal growth points and thus alter the morphology and the size of the crystals [7,8].

Seven different inhibitors were studied in terms of thermal stability, compatibility with the geothermal water, and suitability at the low pH of 6 of the geothermal water. Four of the seven inhibitors passed these investigations and were then tested for efficiency. For this, a model geothermal water was prepared, the inhibitor as well as calcium chloride and sodium hydrocarbonate were added in five steps. The pH was monitored; a continuous decrease of the pH indicates the formation of solid $CaCO_3$. The experiments show that all four inhibitors delay the formation of $CaCO_3$. The largest delay achieves the phosphonate-based inhibitor CUBLEN D 5113.

Partners

We thank the project partners *BWG Geochemische Beratung GmbH, Neubrandenburg, Geothermie Neubrandenburg GmbH, Neubrandenburg, and Geothermie Unterhaching GmbH & Co KG, Unterhaching and Stefanie Wolf, ProScience, Ettlingen.*

References

[1] Hoth, P., Seibt, A., Kellner, T.: *Chemische Charakterisierung der Thermalwässer*, in: P. Hoth, A. Seibt, T. Kellner, E. Huenges (Eds.), *Geowissenschaftliche Bewertungsgrundlagen zur Nutzung hydrogeothermaler Ressourcen in Norddeutschland*, 15, GFZ Publications Potsdam, 1997, pp. 81-89, http://bib.gfz-potsdam.de/pub/str9715/9715-5.pdf.

[2] Monnin, C., Galinier, C.: *The solubility of celestite and barite in electrolyte solutions and natural waters at 25°C: A thermodynamic study,*

[3] Templeton, C.C.: *Solubility of barium sulfate in sodium chloride solutions from 25° to 95°C*, Journal of Chemical & Engineering Data, **5** (1960) 514-516.

[4] Christy, A.G., Putnis, A.: *The kinetics of barite dissolution and precipitation in water and sodium chloride brines at 44-85°C*, Geochimica et Cosmochimica Acta, **57** (1993) 2161-2168.

[5] Canic, T., Baur, S., Kuhn, D., Adelhelm, C., Seibt, A., Wolfgramm, M., Rauppach, K., Möllmann, G.-U.: *Abschlussbericht zum BMU-Vorhaben "Untersuchung der Kinetik von Barytausfällungen aus Geothermalwasser unter dem Einfluss von Scherung und verschiedenen typischen Wärmetauscher-Geometrien" (Förderkennzeichen 0325160)*, 2011, https://www.tib.eu/de/suchen/id/TIB-KAT%3A729929639/.

[6] Canic, T., Baur, S., Bergfeldt, T., Kuhn, D.: *Influences on the Barite Precipitation from Geothermal Brines*, Proceedings World Geothermal Congress, Melbourne, 2015.

[7] Loges, N.: *Kontrolle der Morphologie und Polymorphie von Calciumcarbonat durch monomere und polymere ionische Additive*, Dissertation, Johannes-Gutenberg-Universität, Mainz, 2011.

[8] Reddy, M.M., Hoch, A.R.: *Calcite Crystal Growth Rate Inhibition by Polycarboxylic Acids*, Journal of Colloid and Interface Science, **235** (2001) 365-370.

Read more:

[9] Herfurth, S. (2016). PETher - in situ measurements of thermodynamic properties of geothermal water. GeoTHERM 2016, Offenburg, 15.-16.Februar 2016.

[10] Herfurth, S. (2016). Messung der physikalischen Eigenschaften von Thermalwasser unter Kraftwerks-Bedingungen. GeoTHERM 2016, Offenburg, 15.-16.Februar 2016.

[11] Herfurth, S.; Elisabeth, S.; Thomauske, K.; Herberger, S. (2016). PETher – Physical Properties of Thermal Water under in-situ Conditions. Energy procedia, 97, 240–245. doi:10.1016/j.egypro.2016.10.065

[12] Herfurth, S.; Schröder, E. (2016). On the in-situ measurements of physicochemical properties of geothermal brines in Germany. 26th Goldschmidt Conference, Yokohama, J, June 26 - July 1, 2016.

[13] Herfurth, S.; Schröder, E. (2016). PETher - physical properties of thermal water under in-situ-conditions. European Geo-sciences Union General Assembly, Wien, A, April 17-22, 2016.

[14] Holbein, B.; Isele, J.; Spatafora, L.; Hagenmeyer, V.; Schulenberg, T. (2016). Heat exchanger for down-hole condensation process. 40th Geothermal Research Council Annual Meeting, Sacramento, CA, October 23-26, 2016. Technical Session Papers, 949-955.

[15] Orywall, P.; Dimier, A.; Ukelis, O.; Zorn, R.; Kuhn, D.; Kohl, T. (2016). From experiment to its simulation: An automation of the process. 5th European Seminar on Computing, Plzen, CZ, June 5-10, 2016.

[16] Orywall, P.; Kuhn, D.; Mundhenk, N.; Hillesheimer, D.; Kohl, T. (2016). Induced barite scaling in porous geothermal reservoir rocks during core-flooding experiments. 26th Goldschmidt Conference, Yokohama, J, June 26 - July 1, 2016.

[17] Schulenberg, T. (2016). Entwicklung der Stromerzeugung in Baden-Württemberg und weitere Tendenzen. 93.Mitgliederhauptversammlung des Großabnehmerverbands Energie (GAV), Karlsruhe, 8.November 2016.

[18] Schulenberg, T.; Kuhn, D.; Wiemer, H. J. (2016). Supercritical organic Rankine cycle for geothermal power conversion. 3rd International Meeting of Specialists on Heat Transfer and Fluid Dynamics at Supercritical Pressure (HFSCP2016), Sheffield, GB, August 25-26, 2016.